A LIVING FROM LOB

A Living from Lobsters

R. STEWART

Fishing News Books

© R. Stewart 1971, 1979
Fishing News Books a division of
Blackwell Scientific Publications
Editorial offices:
Osney Mead, Oxford OX2 0EL
25 John Street, London WC1N 2BL
23 Ainslie Place, Edinburgh EH3 6AJ
3 Cambridge Centre, Suite 208
    Cambridge, Massachusetts 02142, USA
54 University Street, Carlton
    Victoria 3053, Australia

First published 1971
Revised edition 1979
Reprinted 1990

Printed in Great Britain by
Billing & Sons Ltd, Worcester

DISTRIBUTORS

Marston Book Services Ltd
PO Box 87
Oxford OX2 0DT
(*Orders*: Tel: 0865 791155
        Fax: 0865 791927
        Telex: 837515)

USA
    Publishers' Business Services
    PO Box 447
    Brookline Village
    Massachusetts 02147
    (*Orders*: Tel: (617) 524–7678)

Canada
    Oxford University Press
    70 Wynford Drive
    Don Mills
    Ontario M3C 1J9
    (*Orders*: Tel: (416) 441–2941)

Australia
    Blackwell Scientific Publications
    (Australia) Pty Ltd
    54 University Street
    Carlton, Victoria 3053
    (*Orders*: Tel: (03) 347–0300)

British Library
Cataloguing in Publication Data

Stewart, Robert
    A Living from Lobsters.—Revised ed.
    1. Lobster fisheries
    I. Title
    639′.54′1   SH380

    ISBN 0 85238 099 2

Contents

List of Illustrations

Foreword

It is possible on any fine weekend to see in most coastal areas which offer reasonable mooring facilities a phenomenal influx of weekend sailors messing about in boats. Many have turned to boats and the sea purely as a form of recreation to fill the longer leisure hours available as a result of the general trend towards a four-day working week, but there are one or two who wish to put their time to more practical use. It is for this type of person—and particularly for the man who decides to go one step further and take up fishing as a full-time occupation—that I have written this book. The advice I give is based on practical experience gained from a lifetime spent in wresting a living from the sea.

I have attempted to give the reader an insight into all aspects of lobstering. I have not glossed over the difficulties and disappointments, for this would be wrong. The life of a lobsterman can be very hard and the newcomer should be prepared for days at sea when the weather is miserable and cold and his pots are consistently empty. On the credit side of course, there are the good times. Times when no other life could be considered as an alternative, for anything else would seem empty and routine by comparison.

One particular point I would make is that equipment which proves successful in a certain area may not be nearly as efficient in another, but I am convinced the creel I have developed and which I describe and recommend is of such a design that it will prove itself in most areas and conditions. It is up to the reader to first carefully consider his chosen area, making sure that potentially good catches are likely, and then to experiment with various types of equipment. It will soon be apparent to the reader that I place a great deal of importance on experimenting. If I get an idea, however much it may depart from traditional thinking, I will mull it over, give it a great deal of thought, and then finally will put it to the practical test. It was in this way that I developed my present creel.

A great deal of thought should also be given to the selection of a boat. In this connection the newcomer must decide whether he is primarily going to use it as a workboat or whether it is to be used more as a pleasure run-about. It is quite likely that he will be able to find one which fills both requirements, but it is far better to have a design which will facilitate easier and safer working rather than one which looks sleek but which in choppy conditions bounces around like a yo-yo.

Admiring glances from the quayside may be very uplifting for your ego, but there will be nobody to admire your courage outside the harbour when it becomes the turn of your stomach to be uplifted!

So my advice is this. By all means take up lobster fishing, but before doing so, carefully consider all the pros and cons of the business. Spend as much time as you can afford amongst established fishermen. Get to know what sort of equipment they use, what sort of boat is preferred for the specific area you have in mind. You may even be lucky enough to meet a fisherman who would be willing to take you on as crew for a few trips. In this way you will discover at first hand what is involved—no amount of reading material can substitute for practical experience.

R. Stewart

Lossiemouth,
Scotland.

Chapter 1

Making a Start

Some years rest easily in the memory so we have little trouble recalling them. The year of 1955 was such a year for me. It was a year when the weather machine seemed to have run out of water, for when the last of the winter snow had melted from the hills during early April we went for seven months without the ground being wet once. High summer temperatures usually bring thunder storms but the sky did not darken once with cloud during the daytime nor were the stars hidden at night. No doubt it was the dryness of the atmosphere which made the air so clear that year. Looking north across quiet water of the Moray Firth, the hills of Sutherland and Caithness had an unusual clarity which even the spring east wind could not dispel. In fact, during the whole of that spring and summer there were long periods when there was a complete calm and the sea lay like a big mirror, not only here in the Firth but all over the North Sea.

During the spring of that year I acquired, for the moderate sum of £20, a small boat some 15 or 16 feet in length which had an inboard engine and was being given up by someone local. She was one of many which had been built in one of the fishing villages that lie on the south coast of the Firth and had been used for line fishing of white fish, haddock, cod and flat fish, before and around the turn of the century. Built on the famous fifie lines for carrying sail, she had a good beam to length ratio. With the sail she could give a good turn of speed despite her beam and did not roll too much with the wind on the beam. In fact, for her length, she was a really good working boat. Clean and sharp forrard, she was good in head seas and, drawing over two feet in water, she would keep her head into wind, lying dead for quite some time before she was drawn off. This is a big asset when one is pulling pots from an aft position because much time is saved in windy conditions.

Although I had had a bout of toying with the idea of going lobster fishing, I did not then have any means of catching them. But this fact did not trouble me during the time my partner and I spent doing a little painting and overhauling the small Harman engine. This task was carried out in the most agreeable outdoor conditions while moored at the end of one of the two basins which make up the harbour. Whether

8

it was the result of our preparations or not the engine gave us not one spot of trouble throughout the season which was just as well—the big sail would have been quite useless to us in the windless days that were to follow.

We had just about completed the work of getting her ready for sea when we heard that the only full time lobster fisherman working the local grounds had decided to give up fishing. What his reasons were we did not know and we did not know what sort of catches he was making. This was probably just as well, for had we known we might not have been so eager to commit ourselves to this method of making a living. Lobster fishing in those days was rather a sometime type of fishing. Unlike seine net fishing where the boats landed their catches at the quayside market to be sold by auction, the lobsterman packed his catch for transport to London or Glasgow. As this operation was usually carried out in the small hold of the boat, it would have taken an inquisitive onlooker to have found out what sort of catches that were being made. Since neither my partner nor I were inquisitive, we were quite ignorant of any catches being made then or in previous years. I just had the feeling that I should go lobster fishing and my partner backed my hunch. It was only later that we found out from those who confine their fishing to the two good fishing months of the year, August and September, that the previous year had provided nothing exceptional in the way of fishing. No doubt poor catches had had some bearing on this fisherman giving up, but we decided to call on him with the idea of purchasing any lobster pots he might have. The result of this visit was fruitful, for we ended up with 20 pots weighted and ready for working, with float ropes and floats attached for all but a few. We were ready to go lobster fishing as soon as we acquired some bait to put in the pots. Not that this was any problem, for at that time there were many fish merchants in the neighbourhood who filleted large catches of haddock, leaving the head and bone which, when doubled over and inserted in the bait string, made an ideal form of bait.

If memory serves me right, we started fishing about the end of the third week in July. As it happened, this turned out to be the ideal time for making a start. We chose a reef which lies one mile to the west of the town and close enough to the shore to be accessible from the sandy beach during very low tides. On the first haul of the 20 pots we had set round this reef, we drew a blank, and put this down to the pots 'singing'. Pots which have soft wood bases tend to make a singing sound when first set as the sea water under pressure is forced into the wood. This usually ceases after 24 hours so when we lifted the first pot early in the morning of the second day we found it contained three lobsters. Hardly able to believe our good fortune we hauled the others and to our delight found that each pot contained at least two or three. In all we had caught 70 odd lobsters from our 20 pots.

Working on the belief that lobster pots should be lifted twice daily (the idea of the second haul being that the bait should be renewed before night-time and any crabs which had entered during the daytime taken out), we set out again. We knew that lobsters, being of nocturnal habit, usually only entered the pots during the hours of darkness, so we were surprised to find that some had departed from this custom and had in fact gone into the pots from the time we had reset them in the morning. Thus we were able to add a further eight to the days catch.

Our haul the following morning found us with more lobsters than we could accommodate in the keep box, for we had again taken about 70 lobsters from the pots. Considering we already had about 150 lobsters and the prospects of a further 70 the next morning we decided to use a tea chest for sending them to Billingsgate. A hunt around the local shops furnished us with half a dozen of these with lids; some of them contained straw which would come in useful when packing the lobsters.

We did not know it then, but from the first time we lifted the lobsters in fish baskets by rope up the side of the quay, lobster fishing was never to be the same again in Lossiemouth. The few lobster fishermen in the area may have been occasionally seen making their way to the railway station carrying a smallish box under their arm once or twice a week when the fishing was at its best, getting little or no response from any who may have witnessed their activities. But eyebrows were raised when we filled our tea chest, putting a layer of straw over each tier of lobsters until the box was full. Eyebrows were to rise still further when we were seen filling another tea chest two mornings later!

We had in fact, either by lucky chance or the application of newcomers' common-sense to a problem, found improved methods of operation which were to stand us in good stead for the future.

Chapter 2

My Kind of Creel

THE methods of fishing for lobsters I describe in this book have been proved over a long period of time and are the results of constant experiment with different types of equipment and ideas. One of the most important points a lobster fisherman should realise is that the creel he uses must be made as attractive to the lobster as possible. A lobster can't be particularly happy when trapped, even though it may get a meal from the bait, and the first natural reaction is for it to tear its way out of prison. I believe in crediting the lobster with a certain amount of intelligence, and my theory is that if the creel is roomy and dark it will be reasonably happy and will settle down.

A disadvantage of a small pot is that if it has two or three inhabitants, they will almost certainly fight, resulting in shed claws and consequent loss of market value. Until a few years ago, I was content to use the normal size creel, that is, 18 in. × 22 in. × 16 in., but subsequently decided to try one with larger dimensions of 20 in. × 24 in. × 18 in., which I found to be more efficient.

I made a direct comparison of the fishing capability of such a creel with that of one measuring 18 in. × 22 in. × 16 in. During the month of April, I dropped 40 large creels and 10 small ones on a ground in about eight fathoms. They were set haphazardly—no particular pattern—with perhaps a large one lying right alongside a smaller one. On the first lifting, I got 17 lobsters and averaged that number each day for the first week, but not one of these was taken in the small creels. The next week, the fishing went down to an average of 12 per day. On the third day of the second week, I decided the experiment had gone on long enough as I still had not taken any lobsters in the small creels, despite the fact that they had been baited in exactly the same way.

Experimenting further, I constructed an even larger pot (42 in. × 24 in. × 18 in.) and this came up to all my expectations by taking lobsters of all sizes. On one occasion, a creel of these dimensions caught 14 pounds of lobster and this convinced me that these crustaceans prefer to enter a roomy creel rather than the more restricted conventional type.

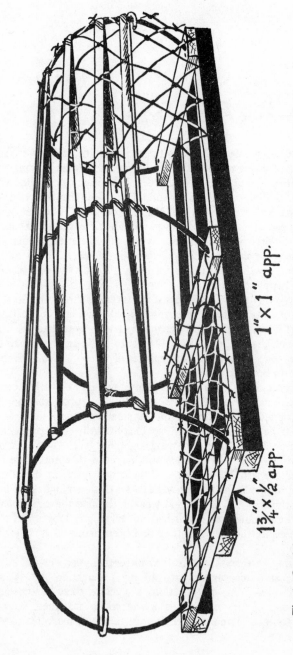

1" × 1" app.

1¾" × ½" app.

Fig. 1. Creel designed by the author. The base is constructed from hardwood, painted black, and forms a light skeleton into which are fixed the hoops of No. 8 gauge galvanised wire. The crossbars are also No. 8 gauge wire. The netting is black nylon and strips of black polythene about 15–20 ft. long are threaded over and under end hoops, through netting and lashed down with twine. This increases the darkness inside the pot. Note the barrel shape of the pot compared to conventional designs.

Providing the ground being worked is productive, say one lobster to every three pots during August and September, I would recommend a trial with this larger size of pot. I am sure it will prove very worthwhile. The only snag is that the larger the creel the more likely it is to be damaged by rough weather, but again I believe I have reduced the risk to the minimum by experimenting and will enlarge upon this aspect later.

My theory that the darker the inside of a pot the more attractive it is to a lobster led me to fix net entrances to ordinary boxes. The results were excellent, but as was to be expected, the boxes were severely damaged by the first spell of bad weather.

I abandoned this idea and turned instead to darkening the inside of conventional creels. Quite obviously the same thing would happen to these if the whole creel was closely covered, so I used black netting rather than green or pink to achieve maximum darkness with minimum resistance to wave action. The best material turned out to be a medium nylon which had not been treated with Cuprinol or similar preservative. Without such treatment this is neither too thick nor too stiff and its attraction to a lobster may be that it looks fragile enough to break through to escape. Again, this idea allows for the lobster having a certain intelligence, and of this I became more certain as time went by.

Digressing slightly, I have observed a form of intelligence by these animals on numerous occasions, but one of the most impressive displays was given by a lobster I brought up in one of my large creels. It was lying at one end of the creel when it came to the surface, but immediately it leapt straight over the two eyes to the other end where a hole had been torn through the netting. I was too quick for him though, and caught him just as he was going over the side. To this day, I swear that he intended doing just that and that it wasn't just coincidence!

But to return to the covering of creels. The light nylon material described does allow one or two lobsters to escape, but I have found that losses are negligible and the use of this material is thus quite justified. By attaching strips of black polythene, one inch wide, to the top of the netting the darkness inside the pot is increased, and painting the base black also helps. Whatever the reason, pots treated in this manner definitely seem to be more productive than others. I must point out that creels covered in this way do tend to lose some of their efficiency the deeper the water gets. It is in shallow water that they excel. The snag here, of course, is that the shallower the water, the more likely it is the pot will be damaged by the motion of the swell.

Lobsters are apparently sheep-like in habit and one can be coaxed to enter a pot if another is already inside. To this end, keep small lobsters which are not of marketable size, tie their claws, and put one into each pot before setting. With two lobsters to a large pot, the latest capture is likely to settle down and not try to break out. Make sure the lobsters you use for this purpose are above legal limits, however. It may be a debatable point whether you are holding an undersized animal or not, and it is better to be on the safe side. The stipulated minimum size of lobsters is 9 in. measured from the tip of the back to the end of the tail when laid as flat as possible.

The type of creel described is ideally suited to areas which have been well fished, that is where production is about one lobster to every five or six pots of conventional design.

Creel entrance

I have no doubt that the entry eye of a creel is of the utmost importance and much thought and care should be given to its construction. A badly designed or sloppy entrance will discourage even the most adventurous lobster. In my experience there are two types of lobster. The first is lively and in size is graded from medium to small, is usually ravenous from August to December due to shell-shedding and is in process of fattening up. The second type is from medium to large. These are older and their barnacle-encrusted bodies indicate that a shell-change has not taken place for a number of years. They are slow and tend to be lazy.

To a certain extent the lively ones are not worried overmuch by the type of entrance. Their main concern is to get at the bait so temptingly on view. It is the larger, lazy types which are harder to capture. They can see the bait, but they aren't going to work to get at it, so the entrance must be made easy and be well-constructed. Ensure that the bottom of the entry eye is not angled as it is in some creels. This may make escape more difficult, but the lazy lobster would rather go without a meal than climb such a slope! The bottom of the eye should be as low as possible on the outside of the creel.

I have found that an eye without a ring—a 'soft eye'—works better than one with this refinement—a further advantage of this type being that it can be attached to the opposite side of the creel, thus keeping the bottom sheet as flat as possible.

Don't be tempted to make the eye from nylon as this has a slippery, insecure surface. It is far better to use heavy cotton or sisal despite the fact that this will have a much shorter life. Also, these materials will not float but will lay flat to the base of the creel when submerged. The eye should be made as large as possible so that the bigger lobster can easily gain access to the pot.

When constructing an eye, I make a semi-circular opening on the side of the creel, about 9 in. wide and 6 in. high and attach the heavy cotton netting to fit this. The other end of the netting is pushed through

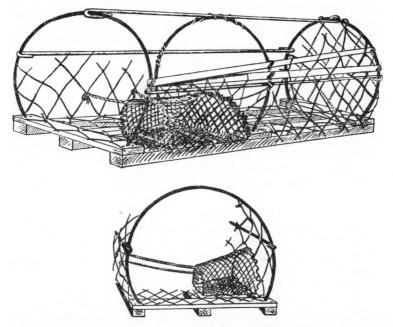

Fig. 2. The eye of the creel should be of heavy cotton or sisal to afford the lobster a good grip, and should be big enough to allow a closed fist to be inserted. Ensure that the bottom of the eye lays flat to the base of the creel and that the inner end is secured to the opposite side of the creel. A larger type of opening specifically designed to attract bigger lobsters can be made in a similiar way but the top should be allowed to hang loose on to the base of the creel. Natural fibre will sink without being weighted and the lobster will easily be able to push through into the creel. With this type of opening, it should be possible to put two closed fists through at the same time.

into the creel and fastened to the base, but with a few meshes left loose and lying flat on the base. Make sure there are no ridges to encumber the lobster's entry.

With a properly designed entrance, you should be able to insert your closed fist through the eye. I have caught many big lobsters using this type of eye—which, incidentally, should not extend more than seven to eight inches inside the creel.

Creels for bad weather

The worst enemy of the fisherman is, of course, the weather, and no matter how much thought goes into the design of a lobster creel from the catching point of view, it is useless unless it can stand up to rough seas. I am a great believer in experimenting with all known methods—

and some that possibly have not been known elsewhere or ever will be! I have had varying degrees of success, but have always learned something whatever the final outcome and my advice is never to allow an idea to remain untried. However unlikely it may seem, it is worth experimenting with.

What happens to the conventional creel with its solid bottom is that it creates so much resistance it is tumbled about until it comes up against a rocky outcrop and either sticks there or keeps hitting against it until the canes break or loosen from the base. It is the frame which suffers damage. Being the lightest part of a creel, it is always the first part which comes into contact with rock and being stiff it will break rather than give slightly with each hammering.

For years I experimented with different designs, trying to get one which would stand up to any weather, but I made the mistake of believing that the heavier and stronger the creel the better it would stay on the bottom. I still lost creels until I learned a lesson from the seaweed. This is neither strong nor stiff, and that was the answer. I would make a creel that had a certain amount of flexibility and which would not move from the place in which it was set.

This meant a light top and a weighted bottom and I came up with the idea of using No. 8 gauge galvanised wire for the frame, which is cut and bent to a hoop-shape and inserted into holes drilled into the base of the creel. Three such hoops are used, and the base I make from hardwood, at least half an inch thick. The three cross bars are cut slightly longer than the length of the base and the ends are formed into U-shapes with a strong pair of pliers. These are then positioned with the U-shaped ends over the two end hoops and nipped tightly with the pliers.

To make a tighter and more durable joint, nip one of the U-shapes in the opposite direction to the other. This will ensure that rough weather will not loosen them.

The cross bars are attached to the middle hoop by a few turns of galvanised wire.

If the creel is given a slight barrel shape in construction, I have found that it will not bend sideways when rolling on the seabed during rough weather. Rather, if it bends at all, it will tend to flatten evenly towards the base, thus retaining its original shape. It is easily pulled back to its proper dimensions, and I find that the covering suffers little or no damage from chafing. The low resistance factor cuts down the risk of the creel driving in heavy swells and thus much time is saved in trying to free it from rocky outcrops or ledges. This is one advantage the conventional wooden-hooped creel does not enjoy.

A great deal of thought should be given to the construction of the weighted bottom. The more weight there is the better it will hug the

16

seabed, but on the other hand this theory falls flat on its face from a handling point of view. The answer is to provide a base with little

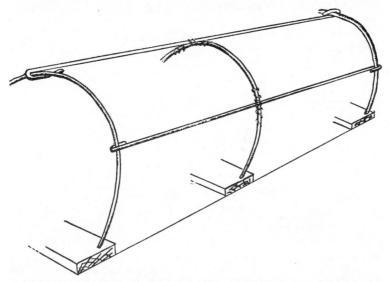

Fig. 3. Fixing the crossbars to the hoops. Turn end joints in opposite directions to give greater rigidity and fix the centre with galvanised wire to the middle hoop.

resistance. The lower the resistance the less the creel will be moved about by the swell.

When making a base, I work towards two things—strength and low bulk. This is achieved by using hardwood which is essential if the wire is going to stay put. If softwood is used, the hole in which the wire is housed will gradually enlarge with movement until the wire simply lifts out and the creel is useless. But even with hardwood, it is necessary to limit the size of the wire, for anything heavier than 8-gauge will both enlarge the hole and chafe the cover. As can be seen from the illustration, the base is made in skeleton fashion. This has two advantages. The first is that netting is cheaper than hardwood and the second is that resistance is kept to a minimum.

For weight in these creels, I use concrete which is reinforced with pieces of galvanised wire running the whole length. The concrete is poured into the middle of the base on to three cross bars which have about six or seven 2½-inch nails driven through them, the protruding ends being bent over. If the concrete is good, it will not work loose from this setting. The total weight is about 17 lb. in small creels and 22 lb. in big ones. Others have complained to me that the nails used to anchor the concrete rapidly corrode and the concrete breaks away. The theory is that something in the water causes this, but I feel the reason is that the sand used in the concrete contains a certain mineral which attacks the metal.

The nails I use are as thin as possible to avoid splitting the wood, and I find that by burring the points they go in easier.

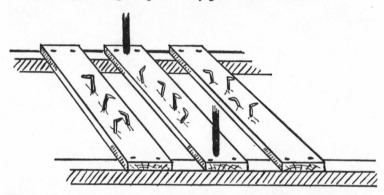

Fig. 4. Concrete is used to weight the creel. Nails driven through the supports are bent over to make a firm anchor. Reinforcement of the concrete is by galvanised wire running through the whole length. About 17–22 lbs weight will be ideal.

I find that the galvanised wire will give about six months' good service before it corrodes and breaks up. After this period of time, I make a point of renewing each creel in turn, each one taking about 30 minutes, but I consider this a small price to pay for the extra dividends one gets from this design. There is, in fact, a plastic covered wire on the market which would probably give even longer life.

Rough weather proves my creels

On one occasion, I had 10 of these creels set on open ground in seven fathoms and these were joined by about the same number from another lobsterman's string. We had both been working this particular spot for about a week when a severe gale came from the WNW. With the wind from this direction there is no swell as the fetch is not long enough, but as the wind was gusting up to about 70 m.p.h., the sea was pretty rough. The waves were high but were not breaking on shore, and when the wind dropped we went out to have a look at the pots. Not one of the other boat's was to be seen nor were they recovered later, but all of mine were still in the same place and I got a few lobsters from them. This well illustrates the ability of my creel design to stand up to rough weather and to resist driving. The other man's creels were wooden-hooped, wooden based and were probably much heavier and set up far more resistance than mine. I cannot stress too strongly the importance of using creels which have minimum resistance. It's not much good producing a low-resistance frame and then covering it with heavy netting. Get a nice light but strong netting similar to the one I have already described.

As I mentioned earlier, losses from this type of creel have been very few, representing savings in both time and money and resulting in good returns. If they are worked singly, I have found that ropes as thin as three-quarters of an inch are quite sufficient, and that a small piece of belting material makes a very good anti-chafer for the float rope. I fix this by nailing it to the base, over the rope.

Chapter 3

Finding a Ground

I WAS always led to believe that lobsters took to deeper waters and went into some sort of hibernation as winter approached. This idea was no doubt due to the fact that fishermen found them very scarce during the winter months, but I have my own theory.

During the summer an area can be most productive, perhaps giving a catch of something like one lobster per creel, but about the middle of October the catch may drop alarmingly to one for every four or five creels. My theory is that the lobsters, rather than taking off for pastures new have simply gone off their food.

The lower temperature of the water in winter makes them less lively and consequently less hungry, but during the summer months, many have to make up extra weight due to shell change in the spring, and solid food is needed for this purpose. By the end of the summer season, the fattening-up process ceases, and they begin to exist more on plankton, which adequately covers their needs during the winter period.

Diet of plankton

In my opinion, the large lazy type which has not had a shell change for a number of years can live out the rest of his life without solid food. Thus, those that were not caught during their fattening-up period may, in theory, escape capture by creel for the rest of their natural lives, but with the constant development of the colony, there will usually be adequate fishing on the same ground for some years.

Possibly the answer to the belief of the older fishermen that lobsters move into deeper water during winter is that those that live in holes close to the shore during winter months tend to come even further in during the summer months and then return to their original holes later in the year. Those that live in deeper water will most likely remain in the same holes throughout the year.

With constant feeding on plankton and no exercise, those lobsters sometimes find they have trapped themselves. While resting, they have put on extra girth and have grown to fit their holes like gloves. Some divers in my area once came across a large one about 14 pounds in weight which had become imprisoned, possibly in this way. The divers reported that the only method of removing him would be with a hammer and chisel! This particular specimen was said to be quite lively despite the fact that he had probably been there for some years without the benefit of solid food.

Skin diver problem

The question of diving for lobsters is a tricky one, but I am not unduly worried about it in this area at any rate. It has been my experience, from what I have seen, that divers can only catch lobsters when there are a lot in the area. I have seen divers operating on a piece of ground and getting few lobsters, and from the same piece of ground at the same time, I have taken quite a few in my creels. So in my view, that is another feather in my cap for the type of creel I use. They catch lobsters even better than skin divers!

Beach fishing

Looking for lobsters on the beach is an engrossing pastime when the sea water one is standing in is comfortably warm. The idea is to probe the dark interior of the hole which usually has a curtain of seaweed and scare any lobsters present into leaving. The lobster will come out more readily if there is a pool of water at the front of the hole.

Surprisingly, the large lobsters with massive claws looking big enough to crush one's hand are the easiest to catch using this method, as their movements are slower than the smaller and younger ones which dart through the water at quite a speed. The lobster is caught by dipping the arm into the pool and gripping the body behind the head. One would be risking a torn finger at least to attempt to grab the lobster from the front—the sharp claws can inflict quite some damage. I recall one fisherman showing me a big scar on his arm the result of an entanglement with a lobster which had severed the main artery. Fortunately, he was able to get medical attention before he lost too much blood.

Fifty or sixty years ago it was possible to make big catches from the beach, but then the catchers took only enough for their needs and for their neighbours. One could find many lobsters in each hole and large crabs hiding under each clump of seaweed, but over the years there has been a steady decline in the stocks of both lobster and crab so that now even the good beach fishers find it hard to get more than one or two.

The big ebb tides of May, June and July are the last times for beach fishing, after this the lobster which had come close inshore would be moving back into deeper water.

Lobster terrain

Lobsters may be found in all parts of the sea bed which provide shelter and it is possible to tell from the lobster's colouring just what type of terrain it inhabited because a lobster adopts the colour most suited to give it camouflage. Thus in rocks which have a heavy covering of long fronds of brown seaweed the lobster will be deep brown-blue in colour and the underside will be brownish white, especially the claws which will be brown underneath. Lobsters living in a more open terrain with less weed will be more white and of a lighter coluring on top, as will those caught from wrecks. The nicest looking lobsters are to be found on flat ledges which carry a covering of sand and very little or no weed at all. These lobsters are mottled with white and light blue patches and the underside is pure white. I found this type of lobster on this sort of ground mainly one mile offshore and separated some distance from heavier ground bearing a lot of weed.

The next question is, where are lobsters most likely to be found? The only answer one can give is to say that they will probably be where there is a rocky bottom or outcrop. The higher and larger the rock the

Fig. 5. Lobster holding ground can vary considerably. Both forms of terrain shown here will be likely to support a colony and should provide good fishing. The top ground, however, will give several years' fishing while the lower ground will be easily exhausted if intensively fished over a period of three to four years. Identification of this type of ground is made by means of a "feeler" (see fig. 6 and text).

more lobsters it will hold, depending, of course, on the amount of fishing it has been subjected to in the past. As I have mentioned previously, the potential of lobster grounds has fallen over the last few years not only in Great Britain but all over the world. More intensive fishing with improved gear will obviously result in greater depletion, but the lobster fisherman has a living to make and he cannot be blamed for getting his share of a good market by using modern methods to the full.

The most obvious solution for any lobsterman is to find a virgin ground. Such a piece of ground will take some finding, but on the other hand it is possible to stumble on it almost by accident. This happened to me some years ago when I was worried by falling catches on my usual ground and began looking for new areas. I had one or two creels set on a rock in seven fathoms of water which I didn't consider a particularly good ground but which gave the odd lobster from one or two pots. This rock was visible from the surface, and I had known of its existence for some years. The dark colouring of seaweed could be seen from the surface, and was very apparent contrasted against the surrounding sandy bottom.

On one occasion, after having lifted the two pots and getting two lobsters, I decided on the spur of the moment to drop six other pots onto a sandy bottom on the way back from the rock. I didn't hold much hope of their being successful as there were no rocks visible which would be likely to house lobsters. The pots settled in a straight line at intervals of 30 to 40 feet. The following day there was quite a heavy swell which prevented me from putting to sea. When I did get out I could not at first find the creels. I took two lobsters from the two pots on the rock and then began searching for the string. I found them, driven by the swell, in a new position about 100 yards away.

On lifting them, I took three lobsters out of each creel, and each one weighed about three pounds. By pure luck and with a little help from the swell, I had found a piece of virgin ground.

I couldn't see any rock in the near vicinity, but nevertheless reset the creels on the same ground. I noted the exact position by means of landmarks, and went to shallower water to pick up my other string of creels which I found badly damaged, they being of the wooden-hooped design I used prior to my successful experiments. With the good fishing I had obtained from the pots on the new ground, I was not unduly worried by this damage, however.

With the arrival of calmer conditions, I went out to take a closer look at the new ground and in the clearer water I could just make out flat rocks on the bottom. There was no seaweed growth on them to make them stand out from the sand and being flat they were virtually invisible.

23

The ground proved to be a very productive one and by trial and error, I found that it stretched some considerable distance from my first discovery. I had 10 weeks' fishing off that area before bad weather put a stop to operations, but returned to it later and fished it regularly for the next two years. I had travelled back and fore across that ground for years and thinking it was just a sandy bottom, had never considered putting a creel on it.

It is apparent from this that it pays to experiment by dropping creels over a wider area, even though, on the face of it, one particular spot doesn't appear to be worthwhile. Also, don't be put off by one bad catch. You will discover, as I did, that the fishing potential of any area will fluctuate from mediocre to very good over a period of time. On one setting on a particular ground, you may find that the catch rate is excellent, but on another occasion, on the same ground, you would declare it useless.

There are several methods of finding lobster grounds, the most obvious being to watch for other boats setting pots. Remember, though, you won't be particularly popular if you set right alongside another fisherman. There is no system of staking claim to an area and no legal method of keeping strangers off your favourite ground, but newcomers have been known to suffer unusual losses and have finally been discouraged. The sensible thing to do of course is to ' read ' the ground and move a short distance away from existing fishermen before setting. At low tide it is possible to see from the beach where there are rocky outcrops, and these same outcrops can also be seen clearly in a depth of 10 fathoms from a boat.

Admiralty charts are also of tremendous help in finding new grounds. They show clearly what sort of terrain the area consists of—whether it is sandy or rocky. Couple this aid with the convenience of a car and you can quickly determine whether an area is likely to be productive. From a high point you can see down into the water for quite some way and should you find a piece of ground which is not too near a heavily populated port, the chances are it has not been too heavily fished and should still carry a decent stock of lobsters. The further you get away from people the more lobsters you are likely to catch.

A swell advantage

It was during my first year of fishing that I learned the advantage of a slight sea swell. One morning on leaving the entrance of the harbour the boat had started to rise and fall. It was the first time she had gone off the even keel since we had started. The swell was coming from the NE and there was no local wind. We had our best catch that morning, putting us closer to reaching our ambition of 100 lobsters from the 20 pots we had set. After counting the lobsters, we became even more eager to achieve this number and decided to haul them a second time after mid-day. We did this and caught as many as we usually did in the evening haul, and still the later haul was as usual. The 100 target was not quite achieved—at the end of the day we were just two short of our goal.

I often wondered why swell helped the fishing. I think it's due to bottom sand being churned and making the water less clear. The lobster is a creature of nocturnal habit and fears the daylight, usually searching for food during the dark. But if it is very hungry, this fear may be overcome when the water is dirty. This was made obvious once again when I was working pots on a strip of ground close to the harbour after a very heavy rain storm which sent tons of dirty brown water down the river nearby and out into the sea. This brown water lying on the surface, got into the area I was working, making it quite impossible for me to see any bottom—much to my annoyance. I was visualising a drop in the catch but in fact it was quite the reverse, for when the evenings catch went up I found I could haul them three times a day obtaining good catches each time. Afterwards, I was sorry to see the water clear up.

'Reading' the bottom

One should get to know local grounds intimately which can be done either by sight or by mechanical methods. By this I mean nothing elaborate, but simply using probing tools such as a length of pipe

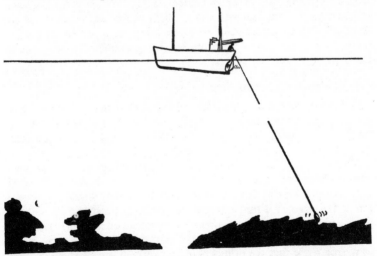

Fig. 6. Using a 'feeler' to identify the type of bottom. Any sort of weight can be used successfully, but a sash cord weight is ideal, and will be less likely to snag fast on the ground shown on the left.

towed from astern. I have found that the balance weight from a window frame is ideal for such a purpose. It is about 8 lb. in weight and is long and slim with a hole at the top, to which can be attached the towing warp, which should be fairly thin to give a good 'reading'. Either courlene or nylon is perfect for the job.

Using this type of feeler needs practice to read accurately the terrain it is being towed over, and to achieve this, make a number of runs over ground with which you are already familiar. By this means you will get to know the feel of sand, stones and rocks. You will find that you can tow this equipment quite fast, and the thinner the twine the faster you will be able to work. The ultimate piece of equipment for this type of work, is an echo sounder, which will find grounds in deeper water but this is an expensive item, the potential benefits of which should be given careful consideration before purchase.

Having found the ground, I place the creels by dropping one over the side and towing it gently until it comes up against the tall rocks which usually house the biggest lobsters (might is right). You will find that towing a wooden based creel will be difficult, for it will tend to float off the bottom. With my design, though, this problem is minimised.

Old wrecks make ideal places for lobsters, but they present problems for the fisherman. A certain amount of care is needed when working wrecks as sometimes spars entangle the creels and gear is lost. I have found it a help when looking at the sea bottom to get the sun shining on one side of the boat while peering down the opposite side aft. The deeper the keel of the boat the better, for this gives a greater shadow which, coupled with the sunlight, improves visibility. If the wreck is in deep water, make sure you bring the creels to the surface slowly, for lobsters will die from something similar to the 'bends' experienced by unwary divers.

The disappearing crab

A strange thing about the Moray Firth area is that no-one has ever gone in seriously for crab fishing and one would suppose that the area would be alive with them, but, in fact, there seem to be very few crabs around.

There are many folk around here who can well remember childhood days spent in crab and lobster hunting along the shoreline. Provided with the large type of fish basket used on boats, they would go down to low water mark and find lobsters and crabs in profusion in the rock pools and under clumps of seaweed. They would often return with their baskets brimming over. But now the story is different. I know by my own experience that the numbers of crabs which can be taken in creels is getting smaller each year. This seems to be pretty general with all types of fish, and whether it can be attributed to pollution, the introduction of the seine net, or lack of feeding, I just don't know. It is something which should be studied and studied seriously if the sea is to continue to provide an important part of the protein needed by the world's population nowadays.

Chapter 4

My Method of Working

Over the years I have noticed that winds always seem to increase in strength as the sun climbs higher into the sky, and it will pay the fisherman to get to sea as soon after daybreak as possible when there are usually two or three hours with little or no wind. Sailing in wind can be a pleasure, but picking up floats and creels is another matter entirely.

Compasses are not used very often in small boats, at least, not in my experience. Usually landmarks are used as a means of fixing the position of good grounds or areas where pots have been set, but there will almost certainly come a time when a compass is essential and one should always be carried on board. Accuracy is, of course, an essential requirement of a compass, and if air bubbles are trapped in the liquid they will affect the reading so should be evacuated. This is done quite simply by removing the stud, which is located on the side of the compass, and adding preferably alcohol, or as a substitute, distilled water.

If I intend working a piece of ground over a long period, I make a habit of knowing the compass direction of it in relation to the harbour, the time taken to get to it coupled with the speed of the engine, and the strength and direction of wind and tide at the time. In foggy weather I am then able to judge almost exactly where my pots are set by re-enacting that run. Of course, if the strength of wind and tide differs from that first run, then this must be taken into account. The longer you have to steam the greater can be the error, so when I have run the full time and still cannot see the corks of my float ropes, I drop a creel to which is attached a float rope and large ball. This gives me a point from which to start searching, and if the fog is not too dense, I can spot the ball from some distance.

I have found that very often fog lifts with the heat of the sun. It may clear completely or it may only thin out a little, but it is worth having a little patience and not putting to sea too early on such mornings. What I like to do is to get to sea immediately there is a sign of the fog lifting,

28

do the necessary work, and be on my way back before it thickens up again which it will often do, particularly if the wind is in the east.

Fog can put you in some pretty awful situations at times, and I recall one episode which could well have turned out to be fatal. I had been looking for a piece of virgin ground for some time and eventually found one that looked promising. I dropped 20 medium-sized creels and knew by the feel of the ground that I would get good fishing from it providing no one else had been there before me. Next morning the fog was pretty thick and I didn't get to sea until about 7.30 a.m. Just before I got to the spot where my creels were set the fog lifted enough to let me see the land, and although it came down again, I found my floats quite easily. The water was very clear, which is usual in foggy conditions, and I could see the creels on the bottom. The first had two good lobsters in it and my hopes ran high, but the second was empty and the side torn where the occupants had escaped. The third creel contained two more good specimens. While I had been hauling, the fog had lifted completely and it was now quite clear. Looking towards the shore, I could see two or three people waving frantically and obviously shouting. It was apparent they wanted me to move away from the area, but as the fishing seemed successful I pretended I couldn't hear them and pointed first to my ears and then to the engine room to indicate this fact.

The next creel contained two lobsters, but the shouting was getting louder and now they had a whistle so I couldn't ignore them any longer. I steered towards them and they informed me that I had no business to be there and should move off immediately. Apparently they used this part of the shoreline as a target range and there was a risk of stray bullets going seawards. While I was in the vicinity they couldn't get on, so I agreed to move off and come back later in the day.

Later on, after making a phone call, I discovered they would be finished with the range at about 4 o'clock in the afternoon. They had every right to be there, for the council had granted them facilities and a bye-law had been drawn up forbidding boats to enter the area when a red flag was flying. As I had not seen this flag on any of my previous visits, I had assumed the place was unused and the fog had prevented me from seeing it on this occasion.

I left the harbour at about 3 p.m. expecting to get a good haul of lobsters from the creels still down, but I was in for a shock. There was not a lobster in sight. Every creel had a breakout. It was obvious that there had been lobsters in each pot, but they had decided not to wait for me. This needed figuring out. The water was very clear and shallow and the sun strong, and when I looked down at the ground every detail could be seen. There was no seaweed around to give shade, and the result was that the whole area was bathed in strong sunlight. This, coupled with the lobster's natural fear of crabs, obviously made the situation unbearable for them and they broke out of the creels. Had I lifted all the pots at daybreak, I am sure I would have got an excellent catch. Also, had I been using my later creels with their black polythene strips, I would probably have fared better.

Spurred on by the size and quality of the lobsters I had taken, I repaired the holes and returned the creels to the bottom. Next morning, I was out early, but not one lobster had entered the pots. I think the reason for this was that those lobsters had gorged themselves on the bait and were not interested in more solid food. It was after this episode that I started attaching the bait by the method I now use. The ground being open, the colony was probably not very large and there were no ravenous youngsters there to go for the bait a second time.

I tried the ground two or three times after this, but I didn't catch any more of those large specimens so I gave up in disgust. The only satisfaction I got from this episode was that it proved my theory that the large, old lobsters can survive for a considerable time once they have gorged solid food, living only on plankton.

Good lobster area

The Lossiemouth area has always been regarded as a pretty good one, but one particular firm that moved in during 1870 thought otherwise. During the first season, they caught 80,000 lb., in the second season 70,000 lb. and in the third 60,000 lb. This apparently wasn't good enough, for they moved to supposedly greener pastures. There have been a number of large specimens caught both here and on the west coast. One old gentleman recalls a lobster of 47½ lb. which he found lying on top of his creel. That must have been a monster, for I've seen a photograph of a 37-pounder and that was pretty big.

How many creels?

The number of creels required to begin lobster fishing depends mainly on three things: how much one can afford; how much time is available for making them and how productive is the area in which fishing will be carried out.

When my partner and I started fishing, we had only 20 or so creels and we caught quite a lot of lobsters. So if you are in an area which has not been overfished, 20 to 30 creels could give reasonably good returns, and more so if the creels are of the type I have described. Although the hardwood used in the base may be expensive, it has a higher resistance to worm than softwood and will pay dividends in longer life. The smaller type of creel described can be made for a low price each. You will require ropes for them and if you are going to work them singly you should allow about eight fathoms per creel, depending of course, on the depth of water at high tide, during which time there should be about a fathom of rope on the surface. I have worked this type of creel with three-quarter-inch circumference polypropylene and found it to be quite strong enough, but a heavier quality should be used if it is intended to work creels in a string. In this case, you will require short pieces of rope about six feet long to tie each creel on the back rope. This should be thinner than the back rope so that if a creel snags, the thinner rope will break first and it will still be possible to retrieve the remainder of the pots.

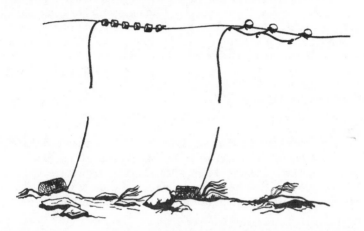

Fig. 7. A good setting. The pots are well in amongst the rocks and the float ropes are of such a length that they make finding and retrieval an easy task. Note the float balls on the right have their own length of anchor rope to guard against losses.

The floats I have always used are of cork, and with this type of rope very few are needed—about six fastened to the top end of each float rope being sufficient. If float balls are preferred, do not tie them directly on to the float rope as in very bad weather these may twist and eventually break the rope with resulting loss of floats. Instead, attach the float ball to the rope with a thinner rope so that this will break first and thus save the remaining floats. Added security can be provided by attaching a few cork floats directly on to the main rope.

I always maintain that ropes of the smallest diameter should be used for attachment to creels, for they offer less resistance in strong tides and heavy swell and therefore not only make the creels more stable but also are not subject to quite so much chafing and subsequent breakage. I prefer to use synthetic rope rather than natural fibre, for the latter tends to rot more quickly.

31

I also found that many of my pots had been lost due to the float ropes being chafed through where they were made fast to the pots. The rope ends are tied to the creel bottoms with a double hitch. The hitch has no protection when the pot starts its inevitable moving with the swell, but I found that I could protect the rope by covering it with strips of thick rubberised fabric belting nailed on to the pots over the rope. This was a big improvement and I hardly lost any more pots through chafing.

Sometimes if the weather was fine and the sea was calm I might see a missing pot on the bottom and catch it with a grapnel, but in fact I was still losing quite a lot of pots. Finally I hit on the idea of weighting my float ropes using a small lead weight. I tied this on two fathoms below the top end of the float rope, the idea being that during the danger period when the tide was low and there was little movement in the water, the rope would be hanging straight up and down rather than having a fathom or so of it lying near the surface. I lost a lot fewer pots after adopting this idea.

Baiting

The matter of baiting is a complex one, and each fisherman has his own ideas, but I find I can make the same piece of bait last much longer than usual. Crabs are a pest when you don't want them, but they can be kept out of the creels. Some time ago, I came to the conclusion after observing the behaviour of lobsters and crabs that the lobster is a

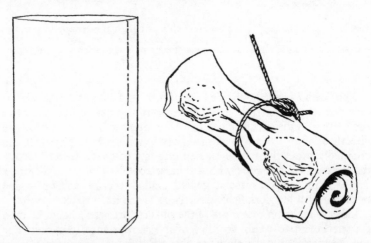

Fig. 8. The plastic bait bag with corners cut to allow circulation of aroma. Because of the slippery surface of the plastic it is almost impossible for the lobster to tear the bag apart and the bait can be used again and again. Also the limited aroma dispersion isn't strong enough to attract crabs.

much smarter fellow. Following along that line of thought, it is possible that his sense of smell is more highly developed, so I tried putting the bait into small tins which had about six holes punched into them. It worked; the crabs kept away from the creels, and lobster catches improved. However, if a crab did manage to get into the pot, it tore the tin to pieces in its efforts to get at the bait. After one or two nasty jags on the rusted metal, I decided to try another method, and experimented instead with plastic bags. A hole no bigger than the diameter of a cigarette is adequate to allow the aroma of the bait to escape and I simply cut the corners from the bag to allow this.

I place the bait in the bag, roll it up and attach it to the bait string in the creel. The crabs still keep away from the pot except in the odd isolated case, and the lobsters are still attracted. It is far easier handling the bags rather than the tins, and storage is not as difficult.

There are two distinct advantages to this type of baiting. The first is that the catch is heavier, and the second is a saving of bait. Prior to using this system, I used up a box of bait on 40 creels, but now the same amount of bait being used over and over again will stretch to as many as 300.

The same basic method, slightly modified, will also suit crab fishing. Strips of plastic about 3 in. × 8 in. are wrapped around the bait, leaving the two ends open to allow circulation of aroma, and the whole package is hung from the bait string in the usual way. The prime advantage of this is that the crab cannot get a grip on the slippery plastic and so the bait is left more or less intact for later use.

To keep bait in a reasonable condition, it is necessary to salt it heavily; an ideal container for this purpose is a dustbin. I have tried most types of bait and in all honesty I cannot say I have found one better than another. I used to believe that mackerel had an edge on other kinds, but I certainly wouldn't swear to that now. Lots of people use shelled mussels, but I have never tried this. I should think it takes quite a time to shuck enough of these to bait a string of pots adequately. If you find yourself without bait, despite using the economic method I have described, a handy tip is to tear up strips of cotton material and dunk them in the brine left in the dustbin. This may sound funny, but it works. Simply insert it in the plastic bag in the same way as flesh bait.

Bait should be easily obtainable from any fishing port. If you have to travel some distance and have to pay for it, follow the advice given regarding the actual baiting of pots and the storing in salt barrels. This will show a considerable saving not only from the financial point of view but also of time spent in going to your supplier.

Setting

Once the ground has been selected, place your creels at the edge of

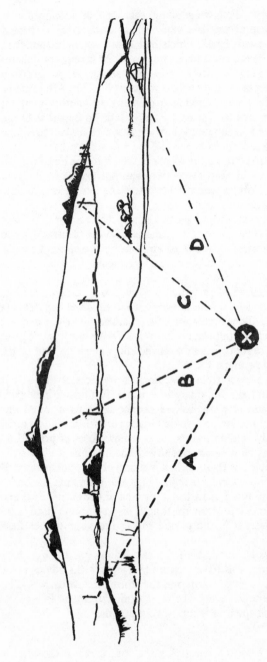

Fig. 9. Having set the pots, it is obviously necessary to be able to return to the same spot to lift them. This can be done by various methods, but the simplest is by using outstanding objects on the coastline as landmarks. It is necessary only to use one or two normally, for once in the approximate setting area the float ropes will be easily seen, but should for any reason the floats not be visible it is possible to start 'creeping' for the pots almost exactly over the area in which they were set. This is done by taking several objects (A, B, C and D) and lining them up in conjunction with each other. For instance, when, as in B, the telegraph pole is directly centred in the fork of the distant hill, then the other objects should all coincide with their individual alignments, confirming that the boat is over the setting area.

the rock outcrop—not on top of it. The largest holes are to be found usually at the base and these will be occupied by the bigger lobsters. With the boat stationary, the first creel should be dropped to the windward side to prevent drifting on to the rope and getting entangled in the propeller. The creel should be lowered until you feel it hit bottom and it should be as close to the rock as possible. On the first pot, leave plenty of float rope until you have determined the exact depth of water. Many pots have been lost because this simple precaution hadn't been taken. Once the depth is known, all other float-ropes can be adjusted accordingly and the pots set with the boat sailing with the wind and tide to ensure that it will always be moving away from ropes. Finally, make sure you can return to the exact spot by taking a fix from landmarks as shown in Fig. 9.

Lifting

The question of lifting pots now arises. How often should they be lifted to provide the most economic catch? I have found that in areas where fishing is good, creels can be lifted twice a day and will give a good return on both settings. In poor areas one lifting will be sufficient.

Conditions play a most important part in lobster fishing, particularly in shallow water areas. I believe that a bit of ground swell is beneficial, providing, of course, that you don't have too much of a good thing! Some of my best catches have been taken in such conditions.

I have a pot hauler on board my boat and use it for hauling from deep water. Even single pots can be difficult when deep, but if you propose working a string of pots a hauler is essential. When I set pots shallow, though, I find it more nuisance than it is worth to use the hauler—it's just as quick to haul by hand. The hard work in this instance is in getting the pots over the side of the vessel and as the hauler only brings them to the surface—which isn't difficult—you actually save time by hauling by hand. Nevertheless, I wouldn't be without this very useful mechanical aid and would recommend anyone taking up lobster fishing to very seriously consider this as an expense to be met.

Fig. 10. A simple tool for retrieving float ropes. The right-angled end is preferable to a hook as, in an emergency, the rope will free itself more easily.

For picking up float ropes, I use a light pole about seven feet in length with a short piece of wood about 12 inches long nailed across one end at right angles. I find this is less likely to be lost than the normal pole with a hook end should a creel snag. The pole with the hook will very likely be pulled out of your hand if you are going ahead at speed.

When picking up creels, steam towards them into the wind and give the boat a bit of helm so as to put slightly away, always taking the creel aboard on the windward side. Whether the creel contains lobsters or not look for holes in the netting, open the door and clean out the creel thoroughly. After rebaiting, tie the door and the creel is ready for dropping again. Remember, *retrieve creels into the wind and set sailing before the wind.*

Don't be discouraged if your pots are empty for the first 24 hours, for new wood 'sings' when put under pressure below the surface This is more pronounced with softwood, and tends to scare off any lobsters within close proximity.

Keeps

One of the problems of a good catch is that in all probability every other lobster fisherman in the country has also had good results, and like any other business a glut means lower prices. To offset this, it is necessary to construct a lobster keep or use floating boxes in which to store the catch until a more remunerative market is available. Obviously, the cheapest method to use is the floating box. This should be constructed of good hardwood and be of sufficient dimensions to allow a large number of lobsters to be kept.

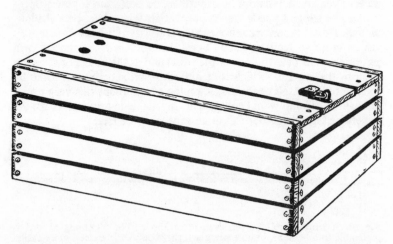

Fig. 11. An ideal lobster keep. The narrow openings allow a constant flow of water while at the same time being too small to enable an eel to get at the lobsters. The lid is kept free of openings as far as possible to prevent pollution from surface debris.

The lobster creel known as a 'parlour pot' is favoured in many areas. This is quite a heavy affair and has the added advantage of being suitable for use as a lobster keep. Being constructed of hardwood slats, it is virtually escape-proof.

My keep is 3 ft. × 2 ft. × 2 ft. 6 in., constructed of good quality hardwood quarter-inch thick by 2 in. wide with quarter-inch gaps between each piece of wood. The size of the gaps is important. They should not be wide enough to allow eels to enter the box and kill the lobsters, and yet should be sufficient to allow a constant circulation of water. I have kept 100 lb. of lobsters in such a box with no losses.

On one occasion, I kept a dozen lobsters in a keep box just to see how they would react to long-term imprisonment. They were left for three months and were still quite lively. I cooked some of them and noticed that the tube which runs down the back was black in each of them, indicating that they had recently been feeding. I was puzzled, for I certainly hadn't put fish in the keep and there were none in the harbour small enough to get into the box, so the conclusion I came to was that they had existed all that time on plankton. I have heard that at the Torry Research Station in Aberdeen lobsters have been kept in boxes for six months without food and all survived. These examples prove conclusively that if only the lobster is handled properly, it can be a most attractive business proposition.

I always suspend my keeps in the harbour rather than in the open sea. No doubt they would get nicer water out there, but the chances of the box being washed away in bad weather are far too great to justify the risk, and I would never recommend this practice. The harbour has its own problems, of course, one of the worst being oil. I have found that by keeping the box well below the surface it will not normally be contaminated, however. Also, by keeping it well down it will not be affected by high temperatures in sunny conditions. None of my boxes has holes in the top and this is for a very good reason. Quite often within a harbour mud is churned up by passing vessels and although this won't do a healthy lobster any harm, it is as well to keep the box as free as possible from falling debris of this sort.

Tying the claws

I think we had been fishing for about one month when we found ourselves again lying moored up at the quayside engaged as usual in the now tedious task of securing the lobsters claws, so that when we put them into the keep box they would be unable to fight with each other, resulting in a loss of claws. For this task we used pieces of string. Holding the sometimes wriggling lobster between our knees, we would get hold of one claw, secure it so it would not open, then do the same with the other claw.

I think it was my partner's idea that rubber bands might do instead of string, so we thought of cutting up bicycle inner tubes into thin bands. It was a rather bemused cycle repairman who appeared from the rear of the shop and presented us with a dozen or so of these which were past using for what they had been made for. The first time we used them we finished the task in about a quarter of the time we had previously taken. After that we were amused to see other lobster fishermen still using string spending a lot of time at the quayside tying up their catch. I believe it was the following year when after seeing the obvious time saved in using bands, the method was adopted by the other fishermen in our area, but it took quite some time for the change to come.

Despatch

Despatch to market is another problem. In too high temperatures and dry conditions, the lobsters will not survive even the shortest of trips. I find that newspapers are cleaner and just as efficient as sawdust or shavings for packing, and to keep the temperature down, I put in a sealed plastic bag containing 8-10 lb. of ice. The lobsters appear quite happy in this environment and survive long journeys. I make a few holes in the top and bottom of the box to allow circulation of cool air.

My original method of packing, though, was anything but successful. Fourteen years ago, when I began fishing with a friend, we were working twenty creels and getting about 60–70 lobsters each day. We packed them in tea chests for despatch to market, but due to high temperatures they were arriving dead and we were getting only 2s.9d. (13·75p) per pound for them. Eventually it dawned on us that they were dying from the heat, and so I decided to experiment. I collected about 30 lb. of ice from the local supplier and scattered this over the lobsters as we packed them. I remember an old fisherman telling me that I would kill them, and I must admit that I rather agreed with him when I saw the steam rising from them as the ice melted. But we had nothing to lose. They were already arriving dead so we couldn't make matters worse. Two or three days later, I got a wire from my agent informing me that this box was the only one he had received in which lobsters were still alive. We received 3s.3d. (16·25p) per pound for them and from that day it was ice in every box we sent off, but packed in plastic bags as already described.

The strange thing about that first summer's fishing was that we rarely had to move the creels. We simply lifted them, removed the lobsters, dropped them back in the same place and got similar results the following day. It was in that year that I learned the advantage of a slight sea swell, for we found that in such conditions we could take 80 lobsters from 20 creels in the first setting, and more from two further settings.

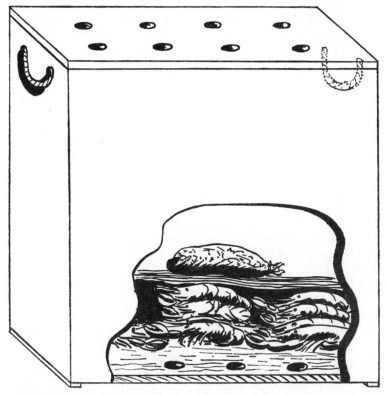

Fig. 12. Good ventilation and low temperature are important requirements when despatching lobsters to market, and this type of box is ideal. It has holes in top and bottom, newspaper packing and 8–10 lb. of ice. It is essential that lobsters are packed in not more than five or six rows deep. Seaweed and green grass should never be used as packing material because of its tendency to generate heat which will kill the lobsters. A box such as this can easily be returned if the lobsterman's name is stencilled on it.

A lesson in transporting lobsters

Young healthy lobsters will survive for quite some time out of water. I once kept one for six days on the deck of my boat without any water at all. I released it, and off it went as lively as ever, but a certain skipper didn't have much luck with an idea he had for transporting lobsters from the west coast to market.

He had a large boat of about 60 ft. and thought he would like a change of fishing. Knowing of the existence of good lobster stocks on the west coast of Scotland he thought he would try his luck there. The problem was one of transporting them from that remote area to market, and he hit upon the idea of installing a live tank in his vessel. He went about it in the right way and provided a circulating pump to constantly supply freshly oxygenated sea water to the captured lobsters. The work completed, they set sail for the grounds and no doubt some of the crew were apprehensive about the financial return of such a venture. They knew from experience that the grounds were profitable, but there was the problem of getting the animals live to market. Losses would probably be heavy, despite the provision of the tank.

The outcome was that they caught a great many lobsters and the tank was soon filled to capacity. They set sail for home and the crew enjoyed the fine weather. Their enjoyment increased as each mile went by and the lobsters were doing well. They were lively and there was no sign of any of them dying. The tank, they believed, was a success.

They returned home via the Caledonian Canal, but before entering it the pump was switched off to prevent fresh water from being taken in. They were about halfway through the canal when one of the crew reported that the lobsters were not as lively as they had been. However, as they would soon be home this didn't worry them too much. Just before leaving the canal however, it was discovered that the lobsters were anything but lively. Several were lifted out of the tank and were found to be dead—in fact, the whole batch had succumbed to lack of oxygen! All the hard work of the trip had been for nothing.

The lesson learned from this was that lobsters kept in water without oxygenation would die quicker than if they were left completely dry. Had the tank aboard that vessel been drained instead of just being deprived of circulation, the story would have had a happier ending. The journey through the canal takes just one day, and they would have survived that with ease.

Marketing

This is, perhaps, the most important part of lobster fishing—or any fishing, for that matter. It is quite useless to be a successful fisherman if you haven't an outlet for your produce. The most obvious answer of course, is to send your catch to an agent in Billingsgate, and if there is a good train or road service from your area there is no problem. This market always has a good demand for lobsters and prices, high now, are continually increasing. There is also the possibility of selling direct to hotels in your area, but you will probably find they need to be guaranteed a certain number and size of lobster each day so that they may be sure of regular supplies and can plan their menus accordingly. If you can guarantee their requirements, however, you will probably obtain a slightly higher price.

A third method of selling is through an exporter who will probably arrange to collect supplies directly from you. Such people are willing to collect from almost anywhere in the country providing, again, that a fair supply can be guaranteed. It would be essential, for both the latter types of transaction, to have a fair-sized keep to maintain adequate stocks.

Working costs

Working about 60 creels from a large engined boat, the running expenses needn't be very high. The boat's insurance should be about £2·50 per £100 annually. Harbour dues in this part of the country are £3 per quarter but will be more further south. If the vessel runs on petrol and is registered as a fishing boat with the Board of Trade, the road tax, which forms a large part of the price you pay for each gallon of petrol, can be reclaimed through your local Customs and Excise Office. You should estimate your average weekly cost of lubricating oil and filter changes. Engine and boat repairs can be expensive both in money and time, but providing you spend a little time regularly in maintenance these should not be particularly bothersome.

Obviously, one cannot generalise about costs, for they vary almost from parish to parish, but with a sensible approach you should be able to keep them to the minimum.

The 'season'

When I first started fishing, I was under the impression that lobsters could only be caught profitably during a short season. Of course, to a certain extent this is true if you are using poor creels which in the winter months will suffer considerable damage. Counting the cost of replacement and comparing it with returns on sales, it is apparent the situation can be anything but remunerative. There are exceptions in some years when it is possible to fish until the end of December, but even in such cases if you can't find a piece of virgin ground to work on after the middle of October, the number of lobsters taken with poor creels would not amount to a marketable proposition.

So it boils down to this. The fishing season is governed by two things: the fishing ability of the creel being used and its ability to withstand bad weather. Thus, I have discovered there is no 'season' in lobster fishing. If you use the right equipment your catch will be profitable throughout the year, providing, of course, the weather allows you to put to sea!

Winter outfit

Although going to sea during the winter months is one of the most uncomfortable activities I know, prices then are high and fishing can be very profitable. To make a success of any business one must take the rough with the smooth and there are many days during the summer months when I wouldn't swap my life for any other.

To make life more bearable during winter, there are one or two precautions that can be taken which will ensure that at least one is warm. I always pay particular attention to my hands. Over a pair of thin cotton gloves I wear a pair of plastic ones, which allows perfect freedom of movement coupled with warmth. The only job I found difficult in these gloves was tying lobster claws, and I must admit I had to dispense with gloves for this operation. All other work, though, was perfectly suited to them, even repairing holes in creels was possible after a little practice.

In order to keep dry during wet weather, fishermen have experimented with a variety of coverings, a common one being the oilskin suit. In fact the term oilskin dates from the time before the war when the only protection one had was a frock made of thin cotton or canvas which had been treated many times with linseed oil and hung up to dry after each coating. These worked pretty well against salt water but rain water always got in at the seams. Since then they have been replaced by plastic and fabric frocks and suits. This was a big improvement.

To cover their heads fishermen wore wide brimmed hats of the same material which went by the name of sou-westers, a pretty fair indication of the prevailing winds at the time they came into use. For footwear it was long leather boots which had to be treated daily with dubbing or fat to keep the water out. Nowadays, rubber boots are essential, but they are naturally cold. To offset this, I made insoles from cardboard to which is glued thin polystyrene sheeting. One has only to lay a hand on this material to realise its thermal quality and used in a pair of boots it is ideal. The snag is that after using it in this way for a short time it compresses and has to be renewed, but the price is reasonable and a few shillings' worth would last a lifetime.

For additional warmth, I always carry a handwarmer which runs for hours on one filling of lighter fuel. It's surprising just how much heat this gives out when carried in a pocket under the oilskin. Two heavy woollen jerseys, heavy woollen trousers and woollen underwear complete the outfit and I find this garb adequate for most conditions. At one time I didn't have so much respect for the dangers of the elements and chose not to wear gloves, but after nearly passing out in one particularly nasty bout of bad weather, I made sure I was well wrapped-up on subsequent trips.

There are other advantages to wearing gloves, of course, such as preventing bones from the bait piercing the skin and turning septic and preventing the hands from becoming calloused from ropework.

Chapter 5

The Boat—Selection and Maintenance

POSSIBLY the most important thing in a fisherman's life is his boat, and he should give a lot of thought to the selection of the type most suited to the work he is engaged in. A good sea boat in the small boat class can usually be judged by its beam which should be about a third of the total length. Being able to judge a good boat is something which can only come with experience, but the average fisherman can give a pretty good verdict on a vessel's capabilities.

When you look to the second-hand market, there will be three main considerations: the sum you are prepared to spend; what is available at that time; what type of boat you are looking for. The latter consideration will be governed by the type of work you will engage in. If you intend working an area away from home, a reasonably big boat with suitable sleeping and living accommodation will be needed. A boat of between 25-30 ft. would be ideal for one or two people both from the living and working viewpoints. A vessel of this size should draw about 4-4½ ft. of water, which will give adequate stability, and it should have a beam of between 9-11 ft.

If there is a choice, what shape is best? You may be faced with a choice of fifie, cruiser or transom sterned boats. Certainly the transom stern makes things easier when the propeller is fouled, but this type tends to take more water on board in a following sea, so for my money, I would settle for the fifie or cruiser stern. Final choice is, of course, governed by personal preference.

Another important point to bear in mind is the strength of the vessel. In some harbours, where traffic is heavy and mooring is at a premium, boats are subjected to quite a lot of battering and squeezing from other boats moored alongside. In bad weather, particularly, the inside vessel will take quite a pounding. The strongest boat is a wooden carvel type with sawn frames which will stand up to more knocks and squeezing than most others.

Before deciding finally on any boat, take a good look at her from the working point of view. Is she going to be easily worked? is there plenty of open deck-space for creels and gear?—does she look a 'comfortable' boat? The more modern type which has the cabin and engine forrard, a large open well and side decks aft, and which is steered from the after end of the cabin has advantages. The creels are more easily stowed with less labour in the working space (this being more noticeable when operating fleets or strings of creels) and all crew members get a certain amount of protection by being in the well, where they are less likely to go overboard in rough weather.

However, where there are advantages there are usually disadvantages to go with them, and one of the worst of a boat of well construction is that it is naturally inclined to retain a lot of water in rough weather. Even in harbour such a boat is vulnerable if moored too close to the sea wall as she will collect anything that comes over the top in stormy conditions. Another disadvantage is the length of propeller shaft needed to connect with the engine which is situated forrard. The longer the shaft the more chance there is of the engine being out of line, thus causing wear on the reduction gear and the tail shaft.

The boat I have is 28 ft. long with a beam of 11 ft. 6 in. and she is fully decked. To make her easier for working, I had a hand rail fitted aft to her quarter which is a big help when working on deck. What type of engine should you look for in a boat of this size? I prefer a diesel engine for dependability, and one of 40-60 hp. will give a pretty good turn of speed. The cost of a boat of this type will of course vary depending on condition.

If you intend making short trips and only during the summer season, a smaller type of boat will be sufficient for your needs—say a 15-25 footer for one or two people. Drawing less water than the larger one, it is possible to get in closer to the rocks with less fear of going aground. A boat of this size will be handier if it is at least partially decked; it should certainly have side decks all round and the engine should be properly housed. Again, the engine I recommend is diesel of about 10-20 hp. An advantage of the smaller boat is that it can easily be taken out of the water once the summer is over. It can then be stored, possibly under cover, and maintenance can be carried out easily and comfortably. It should be possible to buy a boat in this range of light construction with a diesel engine at a relatively reasonable price.

Of course, if you decide to have a boat built you can stipulate what you want and have it built exactly to your own design, and if you have been fishing for some time, you can receive assistance, for the purchase of a new boat, from the White Fish Authority's Grants and Loans Scheme. It is well worth contacting them to get full details of the scheme, for there is the possibility that a sizeable grant would be made towards the cost of the new vessel. Naturally it is wise to 'shop around' for you may pay considerably less for a boat from one builder than from another. Don't however, be tempted to settle for the cheapest. Take into account the experience a yard may have in constructing a particular type of boat—if they specialise, they obviously will produce a reliable, seaworthy vessel at the most economic price.

Having mentioned earlier that lobsters are becoming scarce in some areas, it follows that many fishermen are seeking new fields in which to earn a living and this results in many boats being put on the market. This should have the effect of lowering prices, but in fact, this is not the case, for with the departure of working men in coastal waters there is also a remarkable influx of pleasure-seeking weekenders, most of whom require vessels of one sort or another. This demand tends to keep prices up and in some instances, forces them to an uneconomic level for the working fisherman.

This is an important fact to bear in mind when having a vessel built. If possible, incorporate in the design features that would appeal to a prospective buyer who wishes to convert a working vessel into a leisure boat. This can, of course, only be taken to sensible limits—the boat is first and foremost a working boat and therefore should be basically designed for this purpose. A chat with the builder or designer will result in some useful but unobtrusive additions being made which will stand the boat in good stead should you decide to sell at some future date.

If the boat is to be used for trawling, lining or drifting, the engine will have to be capable of taking the strain. But the bigger the engine the larger the vessel is likely to be and a line operator needs something he can handle comfortably in all conditions. Don't fall for the first good-looking vessel you come across. It is better to spend a little time in seeing as many different types and designs as possible. Go to sea with a fisherman who owns a boat similar to the one you are considering buying. In this way you will see how it handles in practice, what the advantages and disadvantages are, how many pots can be worked from it, and so on.

Converting a ship's lifeboat

One very popular method of obtaining a boat nowadays is to buy a ship's lifeboat and convert it to the type of vessel required, whether it be for pleasure or for working. Some excellent results have been achieved, but it takes a great deal of time and energy to convert such a vessel into a working sea boat. A lifeboat should always have its keel extended and tapered to prevent rolling in rough weather—if this isn't done, you'll wish you had never gone to sea! The tapered keel also helps in beaching a vessel, for it will not list as much as one with an ordinary keel. Fit the extension by means of long bolts which will go through both the

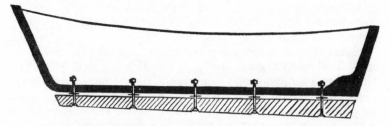

Fig. 13. Extending the keel of a ship's lifeboat. The new keel is attached by means of half-inch cap-headed bolts. Countersunk, these are passed through the extension and existing keel to achieve a watertight fit. A twist of oakum is also used directly below the nuts inside the boat.

extension and the existing keel. These bolts should be placed at about four foot intervals and should be countersunk.

A disadvantage of a converted lifeboat over a certain length is that it will climb over waves rather than cut through them, which means an uncomfortable thump every few minutes in a head sea. This will weaken the vessel after a time unless short hardwood frames are inserted

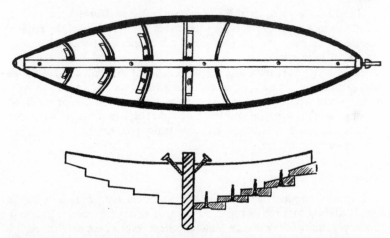

Fig. 14. Fitting lifeboat strengthening frames. These should be constructed of hardwood, such as oak, using a pattern or mould of the boat's planking to achieve a close fit. Countersunk inch-and-a-half brass screws should be used as shown in the lower part of the illustration.

inside the hull. These should be cut so that they form a very close fit from one side to the other and should be secured with heavy brass screws both to the inside keel and to the skin. Boats of this type should never be driven hard into a head sea in any case, so even with the strengthening frames in, this practice should be avoided.

Another advantage of the keel extension is that the boat will lie head into the wind even though the propeller is not turning. This is essential when working creels and makes life very much more comfortable all round.

Always take a friend with you when looking at boats for sale. He needn't be any more of an expert than you, but there's a great deal of truth in the saying that two heads are better than one.

Make a close inspection

I once saw a wooden clinker-built ship's lifeboat for sale, which, at first glance, looked a beautiful craft, but which upon closer inspection turned out to be a complete waste of time. The reason for this was that all the washers behind the copper nails were of galvanised metal and over the years had disintegrated due to electrolysis caused by these two metals being closely related in sea water. The only thing holding that boat together was the shape! The asking price was anything but low, and if anybody did purchase it they bought themselves a lot of trouble. Apart from this type of hazard there is no doubt that a carvel-built boat is stronger and infinitely more preferable than one of clinker construction.

Although wooden boats are usually more lively in the water, those made in synthetic materials have still to be proved as far as I am concerned. They may be easier to maintain and cost less initially, but I know of wooden boats still in use today that were built decades ago from best seasoned wood and which will still be in use for many years to come. Will the new materials last like this? The shape of modern boats in new materials certainly have an edge over those old ones, but whether they ride out heavy seas as well is another matter. I am a confirmed champion of wooden boats, so perhaps I am slightly biased! Some are better sea boats than others, and you may sail in one for months during fine weather in the summer and never find out what sort of vessel you have under you. It is only when you experience rough weather that you can sum up your boat's qualities.

Once you have made your choice, it is almost imperative to have a professional survey made of the vessel. You will be investing a considerable sum of money and it would be foolish not to take such a precaution.

Tide lumps

A good sea boat will stand up to just about anything the weather can throw its way, providing the engine and propeller are in good condition, but there is one type of sea which should always be watched for. This is what fishermen call a 'tide lump'. It is, in fact, a wave of greater height than usual and its width is completely out of proportion to its height. This is what makes it so dangerous, for the angle of its face being much steeper, it hits the boat much higher up and breaks inboard. A large, ordinary wave will not do this because the angle of its face is not nearly so steep. When a tide lump breaks there is more water coming off its crest than with the ordinary type. The Pentland Firth is notorious for its tide lumps as is a short channel of the Isle of Jara, known as the Corrievrechan. Fishermen know that these waters are to be avoided when conditions are bad, but in the open sea one is not usually on the lookout for tide lumps and for this reason they can be very dangerous. Wherever fast tides coupled with winds are encountered, keep a sharp lookout, particularly at periods of full and new moon.

It was a tide lump that caused the loss of the fishing vessel Ocean Swell. This incident was recalled by Billy Clark of Sandhaven, on the occasion of the tragic loss of the lifeboat Duchess of Kent. Writing in the *Sunday Post*, he describes a tide lump:

"It's a combination of a high tide running one way and the gale blowing the other. In water only 40 fathoms deep at best, it's the reason for those 'lumps of water' that fishermen fear. I was afraid that night. The tide was high and running seawards. The gale blowing towards the land. The very setting for a lump. And hit us it did. It was 50 feet high and going like an express train. Roaring like thunder, it caught us broadside. The wheelhouse filled with water. I knew I was drowning. My lungs were bursting. Suddenly there was air. The skipper was slumped unconscious. I realised what had happened. We'd turned turtle. The miracle was the wave had righted us again."

That, then, is a tide lump—a phenomenon which even the sturdiest, most seaworthy of boats finds difficulty in overcoming.

My own vessel is an excellent sea boat, but being on the heavy side, she is not particularly well suited to getting in close to the rocks. I have had her for the last five years and have worked out one or two rules which I put into practice when working close to rocks. I avoid shallow water when the tide is going out or if there is any swell in the sea. A good guide to the depth of water is given by seaweed. If it is just touching the surface at low tide, you can usually be sure of about three feet of water below the surface, but don't put complete faith in this because where the tide runs fast the seaweed will not grow so high and the top of it may be as little as one foot from the seabed. The width of the fronds will give an indication of the height; if narrow, the fronds will be short and *vice versa*.

The echo sounder will not be particularly useful in shallow water either, for the bow can be hard aground on rock and the sounder, which is usually situated aft will be showing six feet of water below the keel. After pushing a heavy boat from rocks once or twice you will not go aground so carelessly in the future! I know, because I've experienced it!

Another of my rules is to keep the anchor in a handy position for immediate use should the engine fail at awkward moments. Close inshore, a creeper will enable you to haul off into deeper water. It is light enough to throw a fair distance and once it has gripped the bottom it will give a good hold even with a strong man hauling a heavy boat towards it. Once in deep water the anchor can be dropped over the side and the emergency is over.

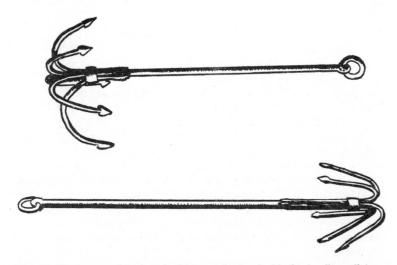

Fig. 15. Two types of 'creeper' which can be invaluable for hauling off into deeper water. The one with the large spread will operate efficiently on either flat or rocky bottoms, but the smaller spread would be more suited to a rocky bottom.

Moorings

If, during the winter, the boat is not used for fishing and is too heavy to take out of the water, ensure that it is in a berth where it won't run back and forth with any swell that may come up. A good berth can be judged by many points, but ease of access is one of the most important. It must be possible to board the vessel without getting soaked or having a long haul in a rowing boat. If she is an open boat, make sure she is not positioned so that when waves break over the harbour wall they fall straight into her.

Plenty of fenders should be provided and these can easily be made from car tyres which can be obtained almost anywhere. Many motorists are only too keen to get rid of them. They should be hung so that the boat doesn't rub against the quay.

Mooring ropes should be of sufficient length to allow the boat to rise and fall with the tide. If another boat moors alongside you, see that its ropes go on to the quay and not on to your boat. This will ensure that your moorings are not taking more than their fair share of the strain. A very good idea which lengthens the life of moorings is to fix a car tyre in the middle of the rope. This is easily attached with ropes going off from each side, one to the quay and the other to the vessel. When strain is applied it is the tyre which takes the brunt of the punishment and not the rope.

Anchor mooring is perfectly adequate for most harbour conditions, but make sure the anchor you use is a good heavy one, and where a rope is used rather than a chain, ensure this is adequately protected at the points at which it is likely to rub. A bicycle tyre is ideally suited to this type of application, for it is light yet sturdy. Simply remove the wire rim, wrap the rubber round and round the rope and secure it. Precautions such as these will lessen the need to turn out in a Force 9-10 gale to check the ropes.

There have been several news reports in recent years of vessels being cast adrift by vandals. To offset any danger of the boat smashing into others and eventually being lost, simply drop an anchor over the side in addition to the usual moorings. Make sure there is enough slack to compensate for the rise and fall of the tide, and your boat, with a little luck, will still be where you left it when you return the following day. Alternatively, a heavy sinker could be used instead of the anchor, the mooring rope attached to this being picked up and made fast as necessary.

Fair weather fishermen

With the approach of the summer season, in any coastal town, you will see hordes of people on the beach on a fine day watching boat-owners prepare their vessels for the work ahead. Paint pots and brushes are out and activity increases as each day goes by. Many holiday makers at a loss for something to fill in the time, ask fisherman if they can accompany them on a trip. If I am approached with such a request, I simply explain that I will be leaving the harbour at about 5.30 the next morning and they will be welcome. Needless to say, not many take up my offer, and I must say I am pleased because I have something of an anti-social streak in my nature. I find taking people out with me a little distracting, especially when they are obviously becoming bored with the trip. Also, I must be in full control of both the boat and any passengers, and many people just can't accept orders. I well remember a party of schoolchildren I agreed to take out in a moment of foolishness. The teacher in charge of this group promised me he would keep them under complete control and that I wouldn't be aware of their presence on board. Well, if what he did was control them, all I can say is that the relationship between teacher and pupil has changed drastically since my days at school! Half a dozen of the more energetic horrors were jumping about all over the boat and I could see some of them ending up as lobster bait. The teacher appeared oblivious of their behaviour and so I cut short that trip and heaved a sigh of relief when they were finally deposited back on terra firma.

Another time I took a family out, and the trip was quite successful until we returned to harbour. Being occupied with docking, I didn't see the woman attempt to disembark on to the stone steps of the quay. The snag was that the boat was still about three feet from those steps, and the silly woman managed to get one foot on them. Of course, the inevitable happened, the boat moved slightly as she tried to push herself on to the steps and she was left doing rather an alarming display of the splits. Her husband grabbed her hand and tried to pull her back on board, but as she was anything but sylph-like, he was not having much success. To prevent a sudden, large displacement of the harbour water, I dropped everything and came to his aid. The two of us managed to get things back to normal and I lectured them on boat procedure for any future trips they might make—with someone else! As a result of these and other events, I repeat, I have an aversion to trippers!

Hull maintenance

With all the painting and scraping going on around and boats looking smarter each day, it's very hard to convince yourself that your boat is all right for another season. But one day the boat in the harbour which looks scruffy and unkempt, you discover with a shock, is yours. So you are shamed into doing something about it.

The first thing to do is beach her; but remember, she's got to be floated off again when ready, so don't, under any circumstances, run her on to the beach at speed, but rather ease her on gently. Choose a flat area and make sure there is nothing on it to penetrate the hull when the tide recedes. I stress a flat beach, because once the boat is high and dry she is likely to settle to any contours and will twist with dire results. Don't select a site at high water mark at the top of the tide. If the boat is heavy, she could be there for a couple of weeks at least before there is another tide high enough to float her off. In this respect, watch out for false high water marks. These occur when there is a deep depression in the Atlantic. The strong winds in the depression push the tide up higher than usual round the British Isles, and I've seen as much as four feet being added to the tide. To put a boat on this mark would lead to a great deal of trouble.

51

If the tides are getting dull, wait until one hour or more after high water before beaching, but if they are making, you can beach your boat at high water without fear of being stranded. Don't beach it, however, if winds are forecast for the area. A heavy swell sweeping up the beach can cause severe damage.

Prior to beaching, lash two tyres together and place these on the bilge side which will come to rest on the beach. To ensure the vessel goes over on to that side, transfer some weight to encourage this as the tide goes out, but make sure everything else is secured to prevent things from falling about once she heels over. As soon as the water is low enough, you can begin scrubbing with a long-handled hard brush. If there is considerable growth, you will find a decorator's scraper useful for removing it. Failing this, dry sand can be used. Stand well back and throw the sand on to the wet bottom. Much of this will adhere and will assist the brushing by forming an abrasive base.

Remove all growth and barnacles. The latter can be unseated with a scraper or knife. Sit back and have a rest—if your boat is big you have earned it by this time! Don't slack though, for, although resting, you can be inspecting the keel to see whether any parts of it are not in contact with the beach. If there are, get pieces of wood and wedges and place these in the appropriate spots. Hammer the wedges one above the other until they are just tight—and restrain yourself from giving it 'one for luck'.

If the boat has been leaking, the places at which the water came in should have been noted, and now is the time to fix them. If in doubt as to the exact place, get some crude oil or paraffin and with a paint brush coat the suspected area from inside. The oil will go through the leak and will be visible from the outside. This should be done while the outside of the hull is still wet. Repair any leaks thus found by using lead or copper sheeting. If copper is used, make little holes around the edges of the patch otherwise you will find that the copper nails will not penetrate it. Incidentally, it is essential to use copper nails with a patch of the same material. Before applying to the hull, put a good coating of tar or a similar substance all over the patch and then tack it on. This will make a good watertight seal which will stand up to considerable wear.

Worm

An area closely covered with small holes will probably be infested with sea worm, but don't be alarmed. To reach a dangerous condition in home waters, the vessel would have to be neglected to a very great degree. Left untreated, the worm will gradually work its way through all the good wood, but treatment is fairly simple. Get a blow lamp and some creosote. Heat the infested area with the lamp, allowing a margin on either side of the visible damage, but don't hold the lamp too close, otherwise the wood surface will burn before the heat has penetrated which is the whole point of the operation. Apply the flame until scorch marks begin to appear and the wood is too hot to touch. If the boat is wet, be even more careful to see that the heat penetrates deeply to ensure thorough drying out. Remove the blow lamp and coat the hull with creosote, which will penetrate right through if the wood is dry and hot. The worms will probably have been killed by the heat, but the creosote will make doubly sure and will give protection against future attacks in that spot.

While the boat is beached, check the propeller. Make sure none of the blades has been damaged or is out of true, and try moving the propeller up and down and from side to side. If there is any movement the stern tube packing gland needs tightening up. If the movement is excessive or if there has been leaking, it is much better to spend a little time on repacking the gland. A slight bend in the tips of the propeller can be remedied without removal of the whole thing. Simply take a heavy flat piece of iron and place this behind the propeller blade while hammering it gently back into shape. If it is necessary to remove the propeller, make absolutely sure when replacing it that you use only brass nuts and split pins to secure it. Electrolysis takes place if two different metals are joined in sea water, and the effect is corrosion which leads to the softer metal breaking down.

It the boat is not too heavy, it is sometimes possible, with a little help, to roll her over on to the opposite bilge so that attention can be given to the other side. Make sure the tyres are in position to receive her, though. If she is too heavy, the only thing to do is to wait until she refloats and then put her on to the other angle.

Painting the hull

A word now about paint techniques. The waterline of my boat is white, and I find that by mixing half gloss and half undercoat I get an excellent result. By adding paint driers to this mixture, runs are avoided to some extent, but it is wise to keep the mixture fairly thick. A waterline that has run into the paint above it and the anti-fouling below looks sloppy and spoils the look of the whole vessel. A much sharper line will be achieved by using an old worn brush rather than a new one.

Always use different brushes for each paint. These are easily cleaned either by using a proprietary brush cleaner or simply turps, petrol or paraffin.

Anti-fouling paint serves two purposes. First, it does what its name implies and keeps the hull free from algae growth and barnacles and secondly it prevents worms from entering the wood. So it is well worth paying for a good one. In fact, where a boat is concerned I believe in getting the best materials. Anti-fouling can, in fact, be thinned slightly by adding turpentine or petrol—neither of which will alter the properties of it in any way.

While painting the hull, pay particular attention to the base of the keel. This part of the boat is usually 'out of sight—out of mind', but quite often gets eaten away by worms and mussels. Make sure it gets a good coating.

Painting the deck

First make sure that the deck is clean and dry. It is a good idea to give it two coats if is subject to a lot of wear. The first coat should be thinned by adding an equal quantity of turpentine, and this will soak right into the deck boards. Let this dry thoroughly before applying the top coat which can be used straight from the can. One very important point to remember when doing the deck is that you have to get ashore—so make certain you are working towards a place from which you can step ashore! Also, remember to close all hatches and so on as you progress towards your point of exit. There are many paints on the market which will give a non-slip surface to the deck, some of which contain ingredients which make the final surface almost like sandpaper. These are ideal, especially if you have 'landlubbers' aboard.

Before painting the deck, of course, all leaks should be stopped by renewing caulking or by replacing deck canvas.

One particularly good caulking material now available is a two-part rubber composition. It provides an incredibly tough and durable seam both to wooden and metal surfaces and dries in about 24 hours. Once applied, it is so tough that it will probably last the lifetime of the vessel. It can be planed or smoothed with a sander, thus providing an absolutely flat deck, and will expand or contract with the deck material. Thus, for an hour or two's work you can have a deck which will need no further attention for years.

If the deck is canvas-covered, and it is necessary to replace part of this, always ensure that the new canvas insert is sealed with an appropriate primer before it is painted.

My own view of painting a vessel is that you can only get out of it what you are going to put into it. Preparation may be tedious, but a well-prepared surface will hold its covering far longer than one that has been skimped. Take your time, make a thorough job of it and you will benefit in the long run.

Chapter 6

Propulsion Hazards—
Prevention and Cure

I HAVE heard on more than one occasion that single creels fish better than a string. Although I have used strings, I mostly work single pots because they can be placed more efficiently than several joined together. Also, they can be moved about more easily without ropes having to be undone. A man who operates on his own will also find single pots an advantage from the handling point of view.

A string of creels can so easily be lost by having its float rope caught in the propeller of a passing vessel and that means far more work in replacement or repair than if a single pot is involved. Should the first hauled pot snag, then it is likely to break the string and a considerable time will be spent in creeping for the remaining pots.

Naturally, in working single pots, each one has its own float rope, and this increases the likelihood of taking rope into the propeller. There are one or two things I do to try and avoid this. I never approach a float rope with the sun in my face and never running with the wind. The latter may seem obvious for the boat will bring up much quicker when the float rope is picked up against the wind, but fouling can easily happen when working a number of pots.

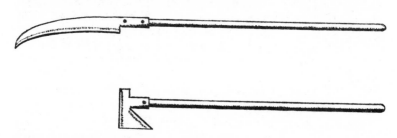

Fig. 16. Cutting tools for freeing jammed ropes from the propeller. With the lower one, it is possible to use a chopping action. A rough file should also be carried to maintain a keen edge on the knives.

Should the propeller become entangled despite these precautions, the situation calls for some pretty quick thinking, particularly if the weather is bad. It is possible for the engine to be stalled by an entangled rope and so the first thing I do is throw the engine into neutral. If thrown into reverse, the outside packing gland may be loosened. I determine how much rope is caught, and if I cannot loosen it by hand —which is very rarely possible—I cut it away by taking hold of the creel end and sawing through. This is reasonably easy if a single creel is being worked, but if a string is involved, the weight of pulling the boat towards it to get some slack can be tremendous, particularly if you have to pull into the wind. The idea of getting some slack is so that you can retrieve the end to prevent loss of the pots.

After cutting and retrieving the creel rope, I determine how much rope has been taken around the propeller shaft. Usually, some has gone round the packing gland, and I cut this away with one of the two long knives I keep on board especially for this purpose. If necessary, you can leave rope around the gland, providing it is not hindering the propeller or has loose ends flapping around. If it hasn't freed itself by the time you return to harbour you can loosen it at leisure. A knife I keep handy has its cutting edge at right angles to the handle so that it can be used in a chopping action to free awkward ropes. Another useful piece of equipment to have on board is a rough file with which to sharpen the cutting edges of knives.

The situation described here is not particularly dangerous, but if a rope is taken into the propeller while the boat is steaming at speed, then the sudden weight of the creel or creels may bend the shaft, resulting in loss of power at best and a ruined reduction gear at the other end of the scale.

The bend may not be great and may be unnoticeable at first glance, but if it is allowed to go untended, the packing gland, reduction gear or gearbox will suffer. There is a simple method of determining whether the shaft is damaged, though, and all you need is a piece of wire long enough to reach from the side of the boat to the shaft about halfway between the stern tube and the shaft coupling. Bend the wire over at one end and fix this with a small nail or drawing pin to the side of the boat in such a way that the other end of the wire will just touch the shaft at 90 deg.

Now, looking down from the vertical, turn the shaft slowly and watch the end of the wire. Any bend in the shaft will immediately show up. This is a much speedier method than removing the shaft.

Ropes are not the only hazard to propellers. One of the worst menaces is discarded or mislaid netting. This is not easily seen in the water but once around the propeller it is the devil's own job to free.

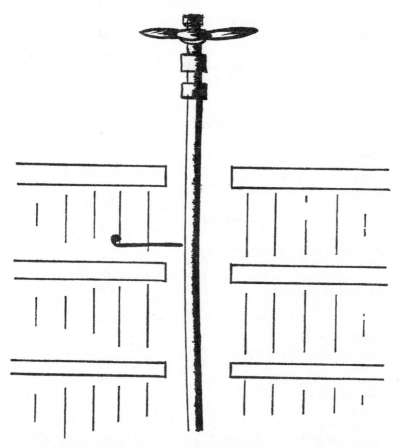

Fig. 17. A simple yet very effective method of determining whether the propeller shaft is out of true. A piece of wire attached alongside and just touching the shaft will immediately show up a bend when the shaft is slowly rotated.

The quicker you throw the engine into neutral the better, for wound tightly it will take considerable effort to free it. A strong piece of wire with a sharp hook on the end is most useful in this situation.

Another menace of this modern age is the plastic bag. It has many practical uses in all walks of life, but round a propeller it just isn't acceptable! Quite often it happens that the boat is steaming along happily when suddenly, with no drop in engine revs the speed falls off. It is rather an alarming state of affairs, and one's first thought is that the propeller itself has dropped off, but on investigation, once the engine has been put in neutral, for you can't see the bag with the propeller turning, the cause is apparent. Some people enclose the propeller with a cage to prevent such flotsam from getting to the propeller. I have never tried this, but in theory it should work.

Another cheap but, I think, essential piece of equipment to carry is a good strong pole of ample length. This you can use to push the boat into deeper water should the propeller or engine fail close in to rocks. And remember to keep the anchor readily available when working pots, especially if the weather is rough and there is a good wind blowing.

A sub-stantial story

Ropes, netting and floating rubbish are not the only hazards a fisherman will come up against in his lifetime. One of the most memorable ones I have encountered was during the second summer of my fishing activities. My partner and I were working our creels inside a long reef, a large part of which breaks the surface of the water. It was a fine morning and the fishing was quite good. Suddenly my partner pointed in the direction of the reef, shouting to me that he had seen something. I looked across, and saw what appeared to be the end of a pole sticking out of the water about 12 inches high. What was so strange about it was the way in which it was moving. It was gathering speed all the time and there was quite a wash coming from it. We judged it to be about three inches in diameter and its speed about 20 knots. It disappeared in a northeasterly direction.

We both discounted the theory that it was some sort of fish, for as far as we knew there wasn't a species with a pole sticking out of its back! We came to the conclusion that it was a submarine and thanked our lucky stars that we weren't nearer! Close to where we were fishing, there is situated a naval air base and we spiced the story with the fact that it was probably a spy submarine having a close look at the base.

Engines

Despite the benefits of the internal combustion engine, many people still consider it to be an infernal invention. They claim that it is the cause of the death of more lobsters and other sea creatures than anything else. By this, I suppose they mean that since its introduction, fishing methods have become more efficient and catches far greater than in the days of sail and oar. It must have taken quite some time to sail or row to where your creels were set and presumably the number of days spent ashore were greater due to bad weather conditions. It took a brave man to face a heavy sea, depending only on the wind or his own muscle power to keep him safe.

However, engines are with us, and will stay. They have improved the lot of the fisherman beyond all recognition, and providing they are looked after reasonably well, will give long life and faithful service.

Before starting any work on the engine, gather around you the spare parts that are likely to need replacing. There is nothing worse than stripping an engine down, finding that a part needs renewing and not being able to get it immediately. A few smooth words to your local garage owner will probably encourage him to supply one or two parts on a sale or return basis. If you find you don't need them, they can more than likely be put in his stockroom.

The first and most important thing to remember about engines is that the fuel, whether it be diesel or petrol, must be clean, so ensure that everything connected with it is treated with respect and care, and always use a filtered funnel when filling the tank. It is no good crying over dirty fuel in rough seas and gale force winds because the carburettor jet is choked or the supply line blocked.

Should the jet become choked, it can usually be cleared by blowing, but in difficult cases it can be opened by means of a very, very thin piece of wire. This should be regarded as a last resort when everything else has failed, for the jet is a precise piece of equipment designed to fine limits. A distorted jet will cause engine failure, or at best faulty running.

In a diesel engine dirty fuel will damage the fuel pump and injectors, again resulting in failure or faulty running—and these are not so easily cleared as a choked petrol jet.

The second thing that goes into an engine is lubricating oil. Use only the best quality and make sure that the oil in the sump is always up to the correct dipstick mark. I make a point of looking at the oil level each time I start the engine. If at any time there is a rise in the amount of oil showing, the presence of water should be suspected. In this case, the sump will have to be drained completely and the point of entry of the water discovered. Water in the engine has the effect of turning oil to sludge with resulting excessive wear on all moving parts.

Ensure that pipes from the fuel tank are securely fastened to prevent vibration and make sure that none rubs against another part while the engine is running. It is no good checking this point if the engine is stationary, for flexible pipes sometimes assume a different shape when oil under pressure is coursing through them. Pay particular attention to electrical wires. Damaged insulation on these can cause shorting and if the wires are adjacent to fuel lines, the result could be disastrous.

Leaking petrol will always find its way to the bilge, so ensure that all connections are sound and pipes are in good condition. If petrol has got into the bilge, clear it at the first opportunity.

Check the sea cock strainer from time to time to ensure that there is a free flow of water. Some of the older types of engine give better results during slow running providing they are kept hot, and this can be achieved by adjusting the flow of water—but take care that sufficient water is getting through to prevent the engine from overheating.

Bilge water should not be allowed to become deep enough to touch the engine or gearbox. Sea water has a strong corrosive action on metal or alloy and will eventually eat its way through.

Magnetos are the cause of many engine failures, and one of the most common reasons is condensation. Keep them as dry as possible and make sure the ventilation hole is free to allow air circulation. Finally, follow the maker's instructions regarding maintenance of an engine and it will give you trouble-free running for many years.

Carburettor trouble

A friend of mine had a petrol-paraffin engine which ran as sweetly as any on fine days, but as soon as a bout of rough weather set in, he experienced trouble—and always just outside the harbour. The engine would stop and when restarted would run for only a very short time before stopping again. The rougher the day, the more trouble the owner could expect. He suspected dirty fuel was the cause, so he took the tank out of the vessel and cleaned it thoroughly. The following day was fine so he took the boat out and she ran beautifully. He complimented himself on a job well done. The fine weather continued for two or three weeks, and the trouble was forgotten, that is, until a wind came up on one trip and the engine stopped, behaving exactly as it had on previous occasions. The fuel and the tank were clean, so obviously he had to look elsewhere. He suspected the carburettor next, so dismantled this, checked the float chamber and jet and found these to be in good order. He cleaned the whole thing thoroughly, reassembled it and put it back on the engine. Exactly the same thing happened—rough sea—engine failure. He was convinced it was something to do with the carburettor, so once more he took it off and went to the local garage for their opinion. Not only the first, but all the garages in town expressed the opinion that it was functioning perfectly.

Not satisfied he took it to the neighbouring town and visited even more garages. Each in turn examined it and finally announced that it was perfect. But at the last garage he was in luck. A representative of the carburettor manufacturer was there and after hearing my friend's story, told him exactly what was wrong.

The whole thing was caused by a worn float pin. In rough seas, this would stick either in the open or shut position, thus starving or flooding the engine with fuel. Used in a car, this type of trouble probably wouldn't have manifested itself, but on a boat with wave action throwing it about it did so only too well! A little piece of five thou. sheeting took up the wear and the carburettor hasn't given a moment's trouble since.

Most engines at some time in their life need expert attention, but many troubles at sea can be put right by just a little knowledge of their method of working. I would consider anybody going to sea without a rudimentary knowledge of mechanics a fool. So much depends on that engine that it is well worth a few hours spent in trying to understand it. Not only is the life of the fisherman endangered by engine failure, but often those of the people who go to his aid. So read the engine manual and become familiar with the elementary procedure one should take to try and restart a failure. There will be times when nothing can be done, but there will be many more when that little bit of knowledge saves a lot of time and trouble.

Chapter 7

Weather Lore

A SUCCESSFUL fisherman always has one eye on the weather. By watching for early signs of storm or rapid change he can have his string of pots safely ashore before they are damaged or destroyed. I can forecast bad weather simply from the appearance of the sky, but this is something one learns over a number of years, and there is always the barometer to interpret weather for beginners.

A ring around the moon is a bad weather sign; the bigger the ring the worse the conditions will be—usually within 48 hours. A 'dirty' coloured sun foretells bad weather within 24 hours and a strong display of Northern Lights means that bad weather can be expected within seven days.

However, complete reliance cannot be placed on such signs. Some storms do occur without prior indication and quite often an area will suffer the effect of a storm centred elsewhere. In my part of the world, the Moray Firth, some of the worst sea conditions are due to storms centred miles away in the Dogger area. The direction of the wind mainly being SE causes a heavy swell to build up and this affects the whole of the Firth. Often such a swell is so great that it breaks miles offshore and may last for days. These conditions can occur without there having been any wind locally, so pots can easily be destroyed without prior warning.

In 1937 the north pier of our harbour was washed away. Despite the fact that this was no flimsy structure, the force of the prolonged swell from the SE simply carried away the half-ton quarry stones supporting the pier. On this occasion, the fishing fleet was out and thus avoided being trapped in the harbour. It was six weeks before the channel entrance was cleared of wreckage and the harbour put to use again.

More northerly winds

We get a greater amount of northerly winds nowadays, and this does not help the fishing on the south side of the Moray Firth, especially during the winter months when they build up to gale force and more.

62

Until 1963 the prevailing winds in this area were southwest and didn't go out much further than northwest, but hardly a week goes by now without our having a gale from the north. When the prevailing winds were southwest we got some hard southeast gales during the winter with the heavy swell which always accompanies these winds.

There is no doubt that the weather pattern is changing and not for the better—especially up here! And not only is the direction of the wind changing, but the amount we get is greater than ever before. I can remember as a child having long spells of what I now know to be anti-cyclone weather, which gave long periods of frosty, but quiet, weather with clear skies and no wind. During these periods the ground seemed always to be covered by snow, but then the weather would break with southeast gales and the heavy swell would lash the coast for days on end, with foam being blown off the sea into the middle of the town. Eventually, the wind would veer into the southwest, the sea would die down and we would return to anti-cyclone conditions.

During the course of a winter there would be four or five of these southeast gales alternating with long quiet periods. You are lucky indeed these days if you can get one week of anti-cyclone weather during the winter. It seems there are no long periods of frosty skies without wind—the longest we seem to get is one or two days of fairly nice weather, and then the wind returns with a vengeance. Should there be clear skies, weather conditions are mild and not crisp and frosty.

Temperatures lower

Summers are also unsettled. Again, in my youth, I remember long hot summers with camping on the beach and plenty of swimming in a warm sea. Many will say that these are memories of childhood which are always good memories, the bad not being logged in a childish mind, but I have proof that conditions have changed, particularly in the temperature of sea water.

I remember one year when I thought winter would never finish. The wind went out to the north about the middle of April and stayed there until the end of July without altering direction, and blowing up to half a gale on most days. The hills across the Firth had a light covering of snow right up until August and when it finally disappeared we thought that at last things were improving. Three weeks later another fall occurred and winter was on its way again!

It wasn't very warm that spring, due to the strong northerly winds, and it was one year when I thought lobsters had disappeared for ever. In previous years I had always made my first worthwhile catches on the 22nd or 23rd of July with amazing regularity, but these dates came and went with no sign of lobster and it was not until the 22nd August that they decided to put in an appearance. This now seems to be the rule, and I am convinced that the sea water temperature is the cause. This had dropped somewhat and the lobsters just will not start feeding until later in the year.

To make up for the lack of lobsters, though, the crabs came inshore in their multitudes that year. I have never seen so many before or since in those shallow waters. The cod fishermen also benefited by the harsh weather conditions, for their fish were driven in by the northerly winds. Although it didn't last long compared with the old days, the fishing was good and profitable. It just goes to show that one man's weather is another man's enemy.

Full moon storms

I believe the worst storms occur at the time of full moon, this theory first having been brought to my notice by a lobster fisherman here who used to accompany his father on salmon fishing trips. He had been told by his father always to look for bad weather at full moon periods, especially early in the year before the weather had settled into the summer months. I have no doubt that the same would go for the end of the year as well, but in their case it probably didn't affect them as they wouldn't work their nets then.

Storms would hit this type of fishing particularly hard, for salmon nets are set in shallow water and are particularly prone to damage from sea swell. In subsequent years, when he was lobster fishing, my acquaintance would take his creels ashore and keep them there for a few days, especially after the summer had finished. This struck me as a waste of good fishing time, particularly as the weather during those periods was not bad, and I made this point to him. It was then that he told me of the full moon theory. From that time on I watched the weather closely during full moon, and I came to realise that there was indeed some truth in his belief.

Something quite different to fishing has also substantiated this theory as far as I am concerned, for it is surprising just how many accidents occur to climbers during the full moon. One reads of people setting out on such trips in the lightest of clothes, the weather being fine and warm, and then a few hours later they are caught in a blizzard and suffer from exposure. Most full moons seem to occur at the weekend and this is just the time when most people are able to indulge in climbing activities.

I have also noticed that if there is a fine spell of weather which continues after the moon has broken (that is, past its full state), it can almost be guaranteed that for the next fortnight at least the weather will be similar. Conversely, if it is stormy before a full moon it will worsen during the full period and will continue unsettled for about a week or so after the moon has broken. That is what I have found in this area.

Reading the signs

Having stated that some storms occur without prior warning, I must now partly contradict myself. There was one particular occasion that sticks in my memory. It was a February morning and I arrived at the harbour at daybreak. Conditions were ideal. There was not a breath of wind, no swell, and if my memory serves me right, the weather forecast for the area was quite good. My boat was moored close to the entrance of the harbour and I noticed that the water in the channel was behaving in rather an unusual manner. Although the sea was like a millpond and there was no wind, the channel was covered with choppy waves three to five inches in height but with no definite direction. It was a condition that I had never encountered before and certainly never since. I don't maintain that these conditions are uncommon—they may well occur at times when I just haven't been around. And in any case, who wants to stand around the entrance to a harbour in bleak weather waiting to see such a happening!

Whatever the reason, I took it to be a sign of approaching bad weather and so put to sea to lift my creels from about seven fathoms of water. It was some little time after returning safely to harbour that the bad weather arrived. A heavy swell rose—without wind—and these conditions continued for about two days. Had I not lifted my gear, the weather would have made short work of it.

I have since read of sensitive instruments which detect such conditions in advance, but the ordinary fisherman cannot avail himself of these facilities, and in any case, the most useful 'equipment' (which is available without charge) can be found on any stretch of coastline between low and high-water mark. It is the winkle!

I place great reliance on these lowly creatures, for I have noticed that they are able to forecast heavy swell conditions, whether there is wind or not, long before such conditions arise. They tend to cling to the rocks in heavy clusters when winter swells are imminent, but in normal weather conditions they will be found well spread out and feeding on seaweed. Of course, I do not suggest that the weather can be foretold accurately by this method, but it is yet another natural indication to be considered.

Studying the weather has saved me a great deal of time and money by enabling me to get my gear in, but at times it is necessary to take risks, although by studying local forecasts these can be kept to a minimum. Life without risks would in any case be dull and uneventful, but on one occasion I decided to leave my gear out and it just didn't work.

About four years ago during the winter I was working one particular ground where fishing was good, but conditions were not particularly brilliant. High winds were the order of the day and I was lucky if I could get to sea on more than two days a week. When I did get out, it was to discover that the pots had taken a good beating and I was faced with repair sessions. However, the design of the pots meant that this would not be unduly tedious, mainly consisting of pulling them back into shape and fixing the odd hole or two. With fresh bait installed they were soon back in the water.

The wind was not going out any further than northwest and although I suffered some damage to the pots, it was not at all severe. But as time passed the wind tended to go a little more to the north which resulted in a heavy swell developing and with force 10–11 gales, conditions were soon bad and damage mounted.

On this particular morning I was up in time for the forecast as usual and this promised nothing better than force 9 northwest. But as the weather glass was quite steady and the sky looked promising, I decided to go out as soon as possible. Steering east-southeast I kept one eye on the weather the whole time.

The string of creels on the ground was very badly damaged and some had disappeared completely. Yes, even my design suffers in atrocious weather. Undaunted, I started to lay more pots but all the time keeping an eye on the sky. Soon there was a darkening and hardening in the direction the wind would come from and as it was very cold I put this down to a snow shower, but a second look convinced me that I was wrong. A snow shower is streaky and this definitely was not so. I decided a hasty retreat was called for and turned the bow towards home. With five minutes to go before entering harbour, the wind hit me from the north and was so strong that I only just made it to my mooring. The entrance to the harbour by that time was white with breakers and an incoming fleet of seiners had to abandon their approach and head for shelter further along the coast.

Lossiemouth is almost impossible to enter once southeast or northeast swells break across the entrance, and there are times when the local fleet is unable to enter for days on end.

With possibly both strings of creels lost in one swoop, this was one risk which didn't pay off, but there was a sequel some time later.

I had decided to give fishing a rest for a while and returned to the same ground in April. Within hours of arriving at the ground, I spotted a float rope just touching the surface during low water. Most of the corks had been torn off and there was just one left to give the rope buoyancy. I started to pull on the rope, not knowing what would emerge, but was completely unprepared for the creel that came to the surface. There was very little damage to it and sitting quite happily in the pot was a lobster weighing about three pounds. He must have been in there for some considerable time.

Tides and swells

It was in this same area that I first appreciated the destructive power of a combination of tide and swell. Although there is very little strength in the ebb tide at this spot, the flood tide runs pretty fast especially during the periods of full and new moons. My estimate is that it runs at about four knots in a south-southeast direction and when this combines with a swell from the northwest or north the full force comes to bear. Destruction of creels at this time is inevitable and losses are great.

With the swell from the southeast cutting across the tide instead of running with it, even though it may be far heavier than that from the north or northwest, I find that damage is negligible.

Obviously the best advice one can give in regard to the weather is to treat it with respect. If you take risks, make sure they are calculated risks and if there is too much doubt in your mind stay ashore. It's better to be safe than sorry!

In Conclusion

So that's the story of lobster fishing. I have tried, and I hope have succeeded, in giving the prospective lobsterman an insight into the business—for make no mistake, it is and must be regarded as a business if you want to make a success of it. It's no good sitting daydreaming on a nice summer's day looking out over a glassy sea and thinking how nice it would be to get away from that stuffy office and simply make money while enjoying yourself 'messing about in boats'. Consider the hard times when you return wet and bedraggled with very little to show for a hard day's work in icy conditions with the sea behaving in a most unlikeable manner. Take into account the long hours you will have to spend repairing and maintaining your boat and gear. Then, if you still have a spark of adventure left, is the time to look to the brighter side. You are your own boss. Whatever you put into the business is for your own benefit.

Any type of fishing of course is a gamble, and the thing to remember is that you will only get out of it what you are prepared to put in. So, once having made up your mind to take the plunge, be prepared to experiment, test new equipment and ideas and generally consider ways in which your methods can be improved. I have found this pays off reasonably well, and I think I have proved my point in these notes.

Some people say the prospects of making a living from lobstering in this country are pretty unattractive. One or two of my predecessors in this area painted a very poor picture of my chances, but in my first year I caught quite a number of lobsters and was encouraged to persevere.

Obviously, there will be a bad year occasionally, but on the other hand there are very good ones during which one can make a good living and if sensible, provide something for the future.

It is just possible that stocks aren't as good as 'in the good old days', but there are often pleasant surprises in this business.

And on that note of optimism I shall end.

Books published by
Fishing News Books

Free catalogue available on request

Advances in fish science and technology
Aquaculture in Taiwan
Aquaculture: principles and practice
Aquaculture training manual
Aquatic weed control
Atlantic salmon: its future
Better angling with simple science
British freshwater fishes
Business management in fisheries and
  aquaculture
Cage aquaculture
Calculations for fishing gear designs
Carp farming
Commercial fishing methods
Control of fish quality
Crab and lobster fishing
The crayfish
Culture of bivalve molluscs
Design of small fishing vessels
Developments in electric fishing
Developments in fisheries research in
  Scotland
Echo sounding and sonar for fishing
The economics of salmon aquaculture
The edible crab and its fishery in British
  waters
Eel culture
Engineering, economics and fisheries
  management
European inland water fish: a multilingual
  catalogue
FAO catalogue of fishing gear designs
FAO catalogue of small scale fishing gear
Fibre ropes for fishing gear
Fish and shellfish farming in coastal waters
Fish catching methods of the world
Fisheries oceanography and ecology
Fisheries of Australia
Fisheries sonar
Fisherman's workbook
Fishermen's handbook
Fishery development experiences
Fishing and stock fluctuations
Fishing boats and their equipment
Fishing boats of the world 1
Fishing boats of the world 2
Fishing boats of the world 3
The fishing cadet's handbook
Fishing ports and markets
Fishing with electricity
Fishing with light

Freezing and irradiation of fish
Freshwater fisheries management
Glossary of UK fishing gear terms
Handbook of trout and salmon diseases
A history of marine fish culture in Europe and
  North America
How to make and set nets
Inland aquaculture development handbook
Intensive fish farming
Introduction to fishery by-products
The law of aquaculture: the law relating to the
  farming of fish and shellfish in Great Britain
The lemon sole
A living from lobsters
The mackerel
Making and managing a trout lake
Managerial effectiveness in fisheries and
  aquaculture
Marine fisheries ecosystem
Marine pollution and sea life
Marketing in fisheries and aquaculture
Mending of fishing nets
Modern deep sea trawling gear
More Scottish fishing craft
Multilingual dictionary of fish and fish
  products
Navigation primer for fishermen
Net work exercises
Netting materials for fishing gear
Ocean forum
Pair trawling and pair seining
Pelagic and semi-pelagic trawling gear
Penaeid shrimps — their biology and
  management
Planning of aquaculture development
Refrigeration of fishing vessels
Salmon and trout farming in Norway
Salmon farming handbook
Scallop and queen fisheries in the British Isles
Seine fishing
Squid jigging from small boats
Stability and trim of fishing vessels and other
  small ships
Study of the sea
Textbook of fish culture
Training fishermen at sea
Trends in fish utilization
Trout farming handbook
Trout farming manual
Tuna fishing with pole and line